DeWALT®

|||||||||||||||||||||||||||||||||||||||||
✓ W9-AEA-532

CONSTRUCTION
PROFESSIONAL REFERENCE

Paul Rosenberg

Created exclusively
for DeWALT by:

PAL
publications®

www.palpublications.com
1-800-246-2175

Titles Available From DeWALT

DeWALT Trade Reference Series

Construction Professional Reference

Datacom Professional Reference

Electric Motor Professional Reference

Electrical Estimating Professional Reference

Electrical Professional Reference

HVAC Professional Reference

Lighting & Maintenance Professional Reference

Plumbing Professional Reference

Referencia profesional sobre la industria eléctrica

Security, Sound & Video Professional Reference

Wiring Diagrams Professional Reference

DeWALT Exam and Certification Series

Electrical Licensing Exam Guide

HVAC Technician Certification Exam Guide

This Book Belongs To:

Name:_____

Company: _____

Title: _____

Department: _____

Company Address: _____

_____

_____

Company Phone: _____

Home Phone: _____

Pal Publications, Inc.
374 Circle of Progress
Pottstown, PA 19464-3810

Copyright © 2005 by Pal Publications
First edition published 2005

ISBN 0-9759709-8-4

09 08 07 06 05   5 4 3 2 1

Printed in the United States of America

A Note To Our Customers

We have manufactured this book to the highest quality standards possible. The cover is made of a flexible, durable and water-resistant material able to withstand the toughest on-the-job conditions. We also utilize the Otabind process which allows this book to lay flatter than traditional paperback books that tend to snap shut while in use.

Preface

Construction is a very complicated set of operations. On a typical project, several thousand different types of materials are required, at least a dozen special skills (usually several dozen), and so many individual decisions that I've never seen anyone attempt a realistic count. On any sizeable construction project, the number has to be hundreds of thousands. On a large project, the number would have to be over a million.

The real purpose of this book is to help you make those decisions. A good decision requires good information. The truth is that there is so much information associated with the construction industry that no one has even a hope of knowing it all. It simply is not humanly possible. Being able to find good information is a major part of the decision-making process. If you can find good information fast, you are efficient. If you cannot, you become inefficient very quickly.

This book contains as much basic construction information as I could fit between its covers. I have attempted to include the most critical and commonly-required information — the information you need to access quickly.

Naturally, a topic may have been overlooked or not covered in depth suitable for all construction workers. I will monitor and update this book continually and will modify it based upon readers' comments and the development of the industry.

Best wishes,
Paul Rosenberg

CONTENTS

CHAPTER 8 – *Painting and Finishing*.. 8-1

CHAPTER 9 – *Plan Symbols* 9-1

CHAPTER 1
General Construction and Safety

DRAWING SCALES			
Plan Use	**Ratio**	**Metric Length**	**English Equivalent (approx.)**
Details:	1:1	1000 mm = 1 m	12" = 1' 0" (full scale)
	1:5	200 mm = 1 m	3" = 1' 0"
	1:10	100 mm = 1 m	$1\frac{1}{2}$" = 1' 0"
	1:20	50 mm = 1 m	$\frac{1}{2}$" = 1' 0"
Floor plans:	1:40	25 mm = 1 m	$\frac{3}{8}$" = 1' 0"
	1:50	20 mm = 1 m	$\frac{1}{4}$" = 1' 0"
Plot plans:	1:80	13.3 mm = 1 m	$\frac{3}{16}$" = 1' 0"
	1:100	12.5 mm = 1 m	$\frac{1}{8}$" = 1' 0"
	1:200	5 mm = 1 m	1" = 20' 0"
Plat plans:	1:500	2 mm = 1 m	1" = 50' 0"
City maps (and larger):	1:1250	0.8 mm = 1 m	1" = 125' 0"
	1:2500	0.4 mm = 1 m	1" = 250' 0"

CSI SPECIFICATION DIVISIONS

Division 1
General Requirements

01010 Summary of Work
01020 Allowances
01025 Measurement and
        Payments
01030 Alternates/Alternatives
01040 Coordination
01060 Workmen's Comp.
        and Insurance
01200 Project Meetings
01300 Submittals/Substitutions
01400 Quality Control
01500 Construction Facilities
01600 Materials and Equipment
01700 Contract Close-out

Division 2
Sitework

02000 Scope of Work
02010 Subsurface Investigation
02100 Site Preparation
02200 Earthwork
02500 Paving and Surfacing
02900 Landscaping

Division 3
Concrete

03000 Scope of Work
03100 Concrete Formwork
03200 Reinforcement
03300 Cast-in-Place Concrete
03400 Precast Concrete

Division 4
Masonry

04000 Scope of Work
04100 Mortar and Grout
04200 Brick
04300 Concrete Masonry Units

Division 5
Metals

05000 Scope of Work
05100 Structural
        Metal Framing
05200 Structural Light
        Gauge Metal Framing

Division 6
Wood and Plastics

06000 Scope of Work
06100 Rough Carpentry
06200 Finish Carpentry

Division 7
Thermal and Moisture Protection

07000 Scope of Work
07100 Waterproofing
07200 Insulation
07300 Roofing Tile
07600 Sheet Metal
07900 Sealants and Caulking

CSI SPECIFICATION DIVISIONS *(cont.)*

Division 8
Doors and Windows

08000 Scope of Work
08100 Metal Doors and Frames
08200 Wood and Plastic Doors
08250 Door Opening Assemblies
08500 Metal Windows

Division 9
Finishes

09000 Scope of Work
09200 Lath and Plaster
09250 Gypsum Board (OWB)
09300 Tile
09500 Acoustical Treatment
09650 Resilient Flooring
09680 Carpet
09900 Painting
09950 Wallcovering

Division 10
Specialties

10000 Scope of Work
10500 Lockers
10800 Toilet and Bath
        Accessories

Division 11
Equipment

11000 Scope of Work
11700 Medical Equipment

Division 12
Furnishings

12000 Scope of Work
12100 Office Furniture
12200 Draperies
12300 Rugs
12400 Art Work

Division 13
Special Construction

13000 Scope of Work
13100 Boiler
13200 Incinerator

Division 14
Conveying Systems

14000 Scope of Work
14100 Elevators
14200 Hoisting Equipment
14300 Conveyors

Division 15
Mechanical

15000 Scope of Work
15050 Basic Mechanical
        Materials and Methods
15100 Heating, Ventilation
        and Air Conditioning
15400 Basic Plumbing
        Materials and Methods
15450 Plumbing

Division 16
Electrical

16000 Scope of Work
16050 Basic Electrical
        Materials and Methods
16400 Service and Distribution
16500 Lighting
16600 Special Systems
16700 Communications

GENERAL CONSTRUCTION CHECKLIST

Excavation
- ☐ Backfilling
- ☐ Clearing the site
- ☐ Compacting
- ☐ Dump fee
- ☐ Equipment rental
- ☐ Equipment transport
- ☐ Establishing new grades
- ☐ General excavation
- ☐ Hauling to dump
- ☐ Pit excavation
- ☐ Pumping
- ☐ Relocating utilities
- ☐ Removing obstructions
- ☐ Shoring
- ☐ Stripping topsoil
- ☐ Trenching

Demolition
- ☐ Cabinet removal
- ☐ Ceiling finish removal
- ☐ Concrete cutting
- ☐ Debris box
- ☐ Door removal
- ☐ Dump fee
- ☐ Dust partition
- ☐ Electrical removal
- ☐ Equipment rental
- ☐ Fixtures removal
- ☐ Flooring removal
- ☐ Framing removal
- ☐ Hauling to dump
- ☐ Masonry removal
- ☐ Plumbing removal
- ☐ Roofing removal
- ☐ Salvage value allowance
- ☐ Siding removal
- ☐ Slab breaking

Demolition *(cont.)*
- ☐ Temporary weather protection
- ☐ Wall finish removal
- ☐ Window removal

Concrete
- ☐ Admixtures
- ☐ Anchors
- ☐ Apron
- ☐ Caps
- ☐ Cement
- ☐ Columns
- ☐ Crushed stone
- ☐ Curbs
- ☐ Curing
- ☐ Drainage
- ☐ Equipment rental
- ☐ Expansion joints
- ☐ Fill
- ☐ Finishing
- ☐ Floating
- ☐ Footings
- ☐ Foundations
- ☐ Grading
- ☐ Gutters
- ☐ Handling
- ☐ Mixing
- ☐ Piers
- ☐ Ready mix
- ☐ Sand
- ☐ Screeds
- ☐ Slabs
- ☐ Stairs
- ☐ Standby time
- ☐ Tamping
- ☐ Topping
- ☐ Vapor barrier
- ☐ Waterproofing

GENERAL CONSTRUCTION CHECKLIST *(cont.)*

Forms
- [ ] Braces
- [ ] Caps
- [ ] Cleaning for reuse
- [ ] Columns
- [ ] Equipment rental
- [ ] Footings
- [ ] Foundations
- [ ] Key joints
- [ ] Layout
- [ ] Nails
- [ ] Piers
- [ ] Salvage value
- [ ] Slab
- [ ] Stair
- [ ] Stakes
- [ ] Ties
- [ ] Walers
- [ ] Wall

Reinforcing
- [ ] Bars
- [ ] Handling
- [ ] Mesh
- [ ] Placing
- [ ] Tying

Masonry
- [ ] Arches
- [ ] Backing
- [ ] Barbecues
- [ ] Cement
- [ ] Chimney
- [ ] Chimney cap
- [ ] Cleaning
- [ ] Clean-out doors
- [ ] Dampers
- [ ] Equipment rental
- [ ] Fireplace

Masonry *(cont.)*
- [ ] Fireplace form
- [ ] Flashing
- [ ] Flue
- [ ] Foundation
- [ ] Glass block
- [ ] Handling
- [ ] Hearths
- [ ] Laying
- [ ] Lime
- [ ] Lintels
- [ ] Mantels
- [ ] Marble
- [ ] Mixing
- [ ] Mortar
- [ ] Paving
- [ ] Piers
- [ ] Reinforcing
- [ ] Repair
- [ ] Repointing
- [ ] Sand
- [ ] Sandblasting
- [ ] Sills
- [ ] Steps
- [ ] Stonework
- [ ] Tile
- [ ] Veneer
- [ ] Vents
- [ ] Wall ties
- [ ] Walls
- [ ] Waterproofing

Rough Carpentry
- [ ] Area walls
- [ ] Backing
- [ ] Beams
- [ ] Blocking
- [ ] Bracing
- [ ] Bridging

GENERAL CONSTRUCTION CHECKLIST (cont.)

Rough Carpentry (cont.)
- ☐ Building paper
- ☐ Columns
- ☐ Cornice
- ☐ Cripples
- ☐ Door frames
- ☐ Dormers
- ☐ Entrance hoods
- ☐ Fascia
- ☐ Fences
- ☐ Flashing
- ☐ Framing clips
- ☐ Furring
- ☐ Girders
- ☐ Gravel stop
- ☐ Grounds
- ☐ Half timber work
- ☐ Hangers
- ☐ Headers
- ☐ Hip jacks
- ☐ Insulation
- ☐ Jack rafters
- ☐ Joists, ceiling
- ☐ Joists, floor
- ☐ Ledgers
- ☐ Nails
- ☐ Outriggers
- ☐ Pier pads
- ☐ Plates
- ☐ Porches
- ☐ Posts
- ☐ Rafters
- ☐ Ribbons
- ☐ Ridges
- ☐ Roof edging
- ☐ Roof trusses
- ☐ Rough frames
- ☐ Rough layout
- ☐ Scaffolding

Rough Carpentry (cont.)
- ☐ Sheathing, roof
- ☐ Sheathing, wall
- ☐ Sills
- ☐ Sleepers
- ☐ Soffit
- ☐ Stairs
- ☐ Straps
- ☐ Strong backs
- ☐ Studs
- ☐ Subfloor
- ☐ Timber connectors
- ☐ Trimmers
- ☐ Valley flashing
- ☐ Valley jacks
- ☐ Vents
- ☐ Window frames

Finish Carpentry
- ☐ Baseboard
- ☐ Bath accessories
- ☐ Belt course
- ☐ Built-ins
- ☐ Cabinets
- ☐ Casings
- ☐ Caulking
- ☐ Ceiling tile
- ☐ Closet doors
- ☐ Closets
- ☐ Corner board
- ☐ Cornice
- ☐ Counter tops
- ☐ Cupolas
- ☐ Door chimes
- ☐ Door hardware
- ☐ Door jambs
- ☐ Door stop
- ☐ Door trim
- ☐ Doors

GENERAL CONSTRUCTION CHECKLIST *(cont.)*

Finish Carpentry *(cont.)*
- ☐ Drywall
- ☐ Entrances
- ☐ Fans
- ☐ Flooring
- ☐ Frames
- ☐ Garage doors
- ☐ Hardware
- ☐ Jambs
- ☐ Linen closets
- ☐ Locksets
- ☐ Louver vents
- ☐ Mail slot
- ☐ Mantels
- ☐ Medicine cabinets
- ☐ Mirrors
- ☐ Molding
- ☐ Nails
- ☐ Paneling
- ☐ Rake
- ☐ Range hood
- ☐ Risers
- ☐ Roofing
- ☐ Room dividers
- ☐ Sash
- ☐ Screen doors
- ☐ Screens
- ☐ Shelving
- ☐ Shutters
- ☐ Siding
- ☐ Sills
- ☐ Sliding doors
- ☐ Stairs
- ☐ Stops
- ☐ Storm doors
- ☐ Threshold
- ☐ Treads
- ☐ Trellis
- ☐ Trim

Finish Carpentry *(cont.)*
- ☐ Vents
- ☐ Wallboard
- ☐ Watertable
- ☐ Window trim
- ☐ Wardrobe closets
- ☐ Weatherstripping
- ☐ Windows

Flooring
- ☐ Adhesive
- ☐ Asphalt tile
- ☐ Carpet
- ☐ Cork tile
- ☐ Flagstone
- ☐ Hardwood
- ☐ Linoleum
- ☐ Marble
- ☐ Nails
- ☐ Pad
- ☐ Rubber tile
- ☐ Seamless vinyl
- ☐ Slate
- ☐ Tack strip
- ☐ Terrazzo
- ☐ Tile
- ☐ Vinyl tile
- ☐ Wood flooring

Plumbing
- ☐ Bathtubs
- ☐ Bar sink
- ☐ Couplings
- ☐ Dishwasher
- ☐ Drain lines
- ☐ Dryers
- ☐ Faucets
- ☐ Fittings
- ☐ Furnace hookup

Plumbing *(cont.)*
- ☐ Garbage disposers
- ☐ Gas service lines
- ☐ Hanging brackets
- ☐ Hardware
- ☐ Laundry trays
- ☐ Lavatories
- ☐ Medicine cabinets
- ☐ Pipe
- ☐ Pumps
- ☐ Septic tank
- ☐ Service sinks
- ☐ Sewer lines
- ☐ Showers
- ☐ Sinks
- ☐ Stack extension
- ☐ Supply lines
- ☐ Tanks
- ☐ Valves
- ☐ Vanity cabinets
- ☐ Vent stacks
- ☐ Washers
- ☐ Waste lines
- ☐ Water closets
- ☐ Water heaters
- ☐ Water meter
- ☐ Water softeners
- ☐ Water tank
- ☐ Water tap

Heating
- ☐ Air conditioning
- ☐ Air return
- ☐ Baseboard
- ☐ Bathroom
- ☐ Blowers
- ☐ Collars
- ☐ Dampers
- ☐ Ducts

Heating *(cont.)*
- ☐ Electric service
- ☐ Furnaces
- ☐ Gas lines
- ☐ Grilles
- ☐ Hot water
- ☐ Infrared
- ☐ Radiant cable
- ☐ Radiators
- ☐ Registers
- ☐ Relocation of system
- ☐ Thermostat
- ☐ Vents
- ☐ Wall units

Roofing
- ☐ Adhesive
- ☐ Asbestos
- ☐ Asphalt shingles
- ☐ Built-up
- ☐ Canvas
- ☐ Caulking
- ☐ Concrete
- ☐ Copper
- ☐ Corrugated
- ☐ Downspouts
- ☐ Felt
- ☐ Fiberglass shingles
- ☐ Flashing
- ☐ Gravel
- ☐ Gutters
- ☐ Gypsum
- ☐ Hip units
- ☐ Insulation
- ☐ Nails
- ☐ Ridge units
- ☐ Roll roofing
- ☐ Scaffolding
- ☐ Shakes

GENERAL CONSTRUCTION CHECKLIST (cont.)

Roofing (cont.)
- ☐ Sheet metal
- ☐ Slate
- ☐ Tile
- ☐ Tin
- ☐ Vents
- ☐ Wood shingles

Sheet Metal
- ☐ Access doors
- ☐ Caulking
- ☐ Downspouts
- ☐ Ducts
- ☐ Flashing
- ☐ Gutters
- ☐ Laundry chutes
- ☐ Roof flashing
- ☐ Valley flashing
- ☐ Vents

Electrical Work
- ☐ Air conditioning
- ☐ Appliance hook-up
- ☐ Bell wiring
- ☐ Cable
- ☐ Ceiling fixtures
- ☐ Circuit breakers
- ☐ Circuit load adequate
- ☐ Clock outlet
- ☐ Conduit
- ☐ Cover plates
- ☐ Dimmers
- ☐ Dishwashers
- ☐ Dryers
- ☐ Fans
- ☐ Fixtures
- ☐ Furnaces
- ☐ Garbage disposers
- ☐ High voltage line

Electrical Work (cont.)
- ☐ Hood hook-up
- ☐ Hook-up
- ☐ Lighting
- ☐ Meter boxes
- ☐ Ovens
- ☐ Panel boards
- ☐ Plug outlets
- ☐ Ranges
- ☐ Receptacles
- ☐ Relocation of existing lines
- ☐ Service entrance
- ☐ Switches
- ☐ Switching
- ☐ Telephone outlets
- ☐ Television wiring
- ☐ Thermostat wiring
- ☐ Transformers
- ☐ Vent fans
- ☐ Wall fixtures
- ☐ Washers
- ☐ Water heaters
- ☐ Wire

Plastering
- ☐ Bases
- ☐ Beads
- ☐ Cement
- ☐ Coloring
- ☐ Cornerite
- ☐ Coves
- ☐ Gypsum
- ☐ Keene's cement
- ☐ Lath
- ☐ Lime
- ☐ Partitions
- ☐ Sand
- ☐ Soffits

GENERAL CONSTRUCTION CHECKLIST *(cont.)*

Painting and Decorating
- [ ] Aluminum paint
- [ ] Cabinets
- [ ] Caulking
- [ ] Ceramic tile
- [ ] Concrete
- [ ] Doors
- [ ] Draperies
- [ ] Filler
- [ ] Finishing
- [ ] Floors
- [ ] Masonry
- [ ] Paperhanging
- [ ] Paste
- [ ] Roof
- [ ] Sandblasting
- [ ] Shingle stain
- [ ] Stucco
- [ ] Wallpaper removal
- [ ] Windows
- [ ] Wood

Glass and Glazing
- [ ] Breakage allowance
- [ ] Crystal
- [ ] Hackout
- [ ] Insulating glass
- [ ] Mirrors
- [ ] Obscure
- [ ] Ornamental
- [ ] Plate
- [ ] Putty
- [ ] Reglaze
- [ ] Window glass

Indirect Costs
- [ ] Barricades
- [ ] Bid bond
- [ ] Builder's risk insurance

Indirect Costs *(cont.)*
- [ ] Building permit fee
- [ ] Business license
- [ ] Cleaning floor
- [ ] Cleaning glass
- [ ] Clean-up
- [ ] Completion bond
- [ ] Debris removal
- [ ] Design fee
- [ ] Equipment floater insurance
- [ ] Equipment rental
- [ ] Estimating fee
- [ ] Expendable tools
- [ ] Field supplies
- [ ] Job phone
- [ ] Job shanty
- [ ] Job signs
- [ ] Liability insurance
- [ ] Maintenance bond
- [ ] Patching after subcontractors
- [ ] Payment bond
- [ ] Plan checking fee
- [ ] Plan cost
- [ ] Protecting adjoining property
- [ ] Protection during construction
- [ ] Removing utilities
- [ ] Repairing damage
- [ ] Sales commission
- [ ] Sales taxes
- [ ] Sewer connection fee
- [ ] State contractor's license
- [ ] Street closing fee
- [ ] Street repair bond
- [ ] Supervision
- [ ] Survey
- [ ] Temporary electrical
- [ ] Temporary fencing
- [ ] Temporary heating
- [ ] Temporary lighting

GENERAL CONSTRUCTION CHECKLIST *(cont.)*

Indirect Costs *(cont.)*
- [ ] Temporary toilets
- [ ] Temporary water
- [ ] Transportation equipment
- [ ] Travel expense
- [ ] Watchman
- [ ] Water meter fee
- [ ] Waxing floors

Administrative Overhead
- [ ] Accounting
- [ ] Advertising
- [ ] Automobiles
- [ ] Depreciation
- [ ] Donations
- [ ] Dues and subscriptions
- [ ] Entertaining
- [ ] Interest
- [ ] Legal fees
- [ ] Licenses and fees
- [ ] Office insurance
- [ ] Office phone
- [ ] Office rent
- [ ] Office salaries
- [ ] Office utilities
- [ ] Pensions
- [ ] Postage
- [ ] Profit sharing
- [ ] Repairs
- [ ] Small tools
- [ ] Taxes
- [ ] Uncollectible accounts

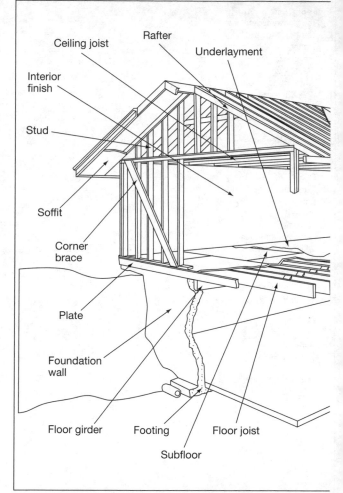

Ceiling joist

Rafter

Underlayment

Interior finish

Stud

Soffit

Corner brace

Plate

Foundation wall

Floor girder

Footing

Floor joist

Subfloor

COMPONENTS

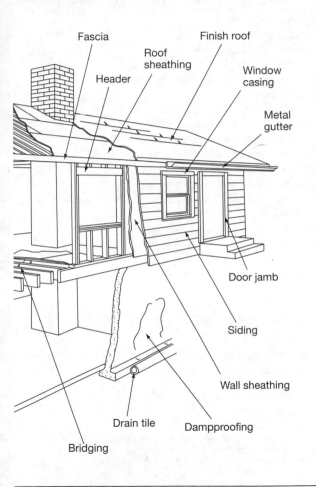

Fascia

Finish roof

Roof sheathing

Header

Window casing

Metal gutter

Door jamb

Siding

Wall sheathing

Drain tile

Dampproofing

Bridging

1-13

INSULATION VALUES
FOR COMMON BUILDING MATERIALS

Exterior Materials

Wood bevel siding, ½ x 8, lapped..............................R-0.81

Wood bevel siding, ¾ x 10, lapped............................R-1.05

Wood siding shingles, 16", 7½" exposure R-0.87

Aluminum or Steel, over sheathing, hollow-backed ...R-0.61

Stucco, per inch...R-0.20

Building paper..R-0.06

½" nail-base insulating board sheathing....................R-1.14

½" insulating board sheathing, regular density...........R-1.32

$^{25}/_{32}$" insulating board sheathing, regular density.......R-2.04

Insulating-board backed nominal ⅜"...........................R-1.82

Insulating-board backed nominal ⅜" foil backed R-2.96

Plywood ¼" R-0.31

Plywood ⅜" R-0.47

Plywood ½" R-0.62

Plywood ⅝" R-0.78

Hardboard ¼" R-0.18

Hardboard, medium density siding $^{7}/_{16}$" R-0.67

Softwood board, fir pine and similar softwoods

   ¾"..R-0.94

   1½"..R-1.89

   2½"..R-3.12

   3½"..R-4.35

Gypsum board ½" R-0.45

Gypsum board ⅝" R-0.56

Temperature Correction Factor
Correction Factor is an ASHRAE standard to be applied for varying
outdoor design temperatures. As follows:

If design temperature is:	-20	-10	0	+10	+20
Then correction factor is:	0.778	0.875	1.0	1.167	1.40

INSULATION VALUES
FOR COMMON BUILDING MATERIALS *(cont.)*

Masonry Materials

Concrete blocks, three oval cores

    Cinder aggregate, 4" thick..............................R-1.11

    Cinder aggregate, 12" thick............................R-1.89

    Cinder aggregate, 8" thick..............................R-1.72

    Sand and gravel aggregate, 8" thick R-1.11

    Sand and gravel aggregate, 12" thick R-1.28

    Lightweight aggregate

    (expanded clay, shale, slag, pumice, etc.),

    8" thick R-2.00

Concrete blocks, two rectangular cores

    Sand and gravel aggregate, 8" thick R-1.04

    Lightweight aggregate, 8" thick.......................R-2.18

Common brick, per inch R-0.20

Face brick, per inch R-0.11

Sand-and-gravel concrete, per inch R-0.08

Insulation

Fiberglass 2" thick...R-7.00

Fiberglass 3½" thick..R-11.00

Fiberglass 6" thick...R-19.00

Fiberglass 12" thick..R-38.00

Styrofoam Board ¾" thick....................................R-4.05

Styrofoam Board 1" tongue and groove.......................R-5.40

Roofing

Asphalt shingles..R-0.44

Wood shingles, plain and plastic film faced R-0.94

Temperature Correction Factor
Correction Factor is an ASHRAE standard to be applied for varying
outdoor design temperatures. As follows:

If design temperature is:	-20	-10	0	+10	+20
Then correction factor is:	0.778	0.875	1.0	1.167	1.40

INSULATION VALUES
FOR COMMON BUILDING MATERIALS *(cont.)*

Surface Air Films

Inside, still air

Heat flow UP (through horizontal surface)
- Non-reflective..R-0.61
- Reflective R-1.32

Heat flow DOWN (through horizontal surface)
- Non-reflective..R-0.92
- Reflective R-4.55

Heat flow HORIZONTAL (through vertical surface)
- Non-reflective..R-0.68

Outside

Heat flow any direction, surface any position
- 15 mph wind (winter) R-0.17
- 7.5 mph wind (summer) R-0.25

Glass

U-Values...Glass Only (winter)

Single-pane glass..1.16

Double-pane ⅝" insulating glass (¼" air space)................58

Double-pane xi insulating glass.......................................55

Double-pane 1 insulating glass (½" air space) 49

Double-panel xi insulating glass with combination
   (2" air space)...35

Temperature Correction Factor

Correction Factor is an ASHRAE standard to be applied for varying outdoor design temperatures. As follows:

If design temperature is:	-20	-10	0	+10	+20
Then correction factor is:	0.778	0.875	1.0	1.167	1.40

THERMAL AND SOUND INSULATION

Material	Kind	Insulation Value
Masonry	Concrete, sand, and gravel, 1"	R-0.08
	Concrete blocks (three core)	
	Sand and gravel aggregate, 4"	R-0.71
	Sand and gravel aggregate, 8"	R-1.11
	Lightweight aggregate, 4"	R-1.50
	Lightweight aggregate, 8"	R-2.00
	Brick	
	Face, 4"	R-0.44
	Common, 4"	R-0.80
	Stone, lime, sand, 1"	R-0.08
	Stucco, 1"	R-0.20
Wood	Fir, pine, other softwoods, ¾"	R-0.94
	Fir, pine, other softwoods, 1½"	R-1.89
	Fir, pine, other softwoods, 3½"	R-4.35
	Maple, oak, other hardwoods, 1"	R-0.91
Manufactured Wood Products	Plywood, softwood, ¼"	R-0.31
	Plywood, softwood, ½"	R-0.62
	Plywood, softwood, ⅝"	R-0.78
	Plywood, softwood, ¾"	R-0.93
	Hardboard, tempered, ¼"	R-0.25
	Hardboard, underlayment, ¼"	R-0.31
	Particleboard, underlayment, ⅝"	R-0.82
	Mineral fiber, ¼"	R-0.21
	Gypsum board, ½"	R-0.45
	Gypsum board, ⅝"	R-0.56
	Insulation board sheathing, ¼"	R-1.32
	Insulation board sheathing, $^{25}/_{32}$"	R-2.06

Temperature Correction Factor
Correction Factor is an ASHRAE standard to be applied for varying outdoor design temperatures. As follows:

If design temperature is:	-20	-10	0	+10	+20
Then correction factor is:	0.778	0.875	1.0	1.167	1.40

THERMAL AND SOUND INSULATION *(cont.)*

Material	Kind	Insulation Value
Siding and Roofing	Building paper, permeable felt, 15 lb.	R-0.06
	Wood bevel siding, ½"	R-0.81
	Wood bevel siding, ¾"	R-1.05
	Aluminum, hollow-back siding	R-0.61
	Wood siding shingles, 7½" exp.	R-0.87
	Wood roofing shingles, standard	R-0.94
	Asphalt roofing shingles	R-0.44
Insulation	Cellular or foam glass, 1"	R-2.50
	Glass fiber, batt, 1"	R-3.13
	Expanded perlite, 1"	R-2.78
	Expanded polystyrene bead board, 1"	R-3.85
	Expanded polystyrene extruded smooth, 1"	R-5.00
	Expanded polyurethane, 1"	R-7.00
	Mineral fiber with binder, 1"	R-3.45
Inside Finish	Cement plaster, sand aggregate, 1"	R-0.20
	Gypsum plaster, lightweight aggregate, ½"	R-0.32
	Hardwood finished floor, ¾"	R-0.68
	Vinyl floor, ⅛"	R-0.05
	Carpet and fibrous pad	R-2.08

Temperature Correction Factor
Correction Factor is an ASHRAE standard to be applied for varying outdoor design temperatures. As follows:

If design temperature is:	-20	-10	0	+10	+20
Then correction factor is:	0.778	0.875	1.0	1.167	1.40

TYPES OF WALL CONSTRUCTION AND THEIR R-VALUES

Uninsulated 2 x 4 stud wall

Air films	R = 0.9
¾" wood exterior siding	1.0
½" insulation board	1.2
Air space	1.2
Vapor barrier	0
½" gypsum board	0.5
	4.8 total R

2 x 4 stud wall with batt insulation

Air films	R = 0.9
¾" wood exterior siding	1.0
½" insulation board	1.2
3½" batt or blanket insulation	11.0
Vapor barrier	0
½" gypsum board	0.5
	14.6 total R

2 x 4 stud wall with rigid board

Air films	R = 0.9
¾" wood exterior siding	1.0
1" polystyrene rigid board	5.0
3½" insulation blanket	11.0
Vapor barrier	0
½" gypsum board	0.5
	18.4 total R

TYPES OF WALL CONSTRUCTION
AND THEIR R-VALUES (cont.)

Improved insulated 2 x 4 stud wall

Air films...R = 0.9

¾" wood exterior siding 1.0

¾" insulation board 2.0

3⅝" batt insulation 13.0

Vapor barrier...0

⅝" urethane insulation board 5.0

½" gypsum board..0.5

22.4 total R

2 x 6 insulated stud wall

Air films...R = 0.9

¾" wood exterior siding 1.0

²⁵⁄₃₂" insulation board 1.9

5½" insulating blanket...19.0

Vapor barrier...0

½" gypsum board..0.5

23.3 total R

Improved 2 x 6 insulated stud wall

Air films...R = 0.9

¾" wood exterior siding 1.0

¾" insulation board 2.0

5½" batt insulation 19.0

Vapor barrier...0

⅝" urethane insulation 5.0

⅝" gypsum board..0.6

28.5 total R

NAIL SIZES

2d: 2 pennyweight x 1" long	16d: 16 pennyweight x 3" long
4d: 4 pennyweight x 1¼" long	20d: 20 pennyweight x 3½" long
6d: 6 pennyweight x 1½" long	straw nail = 10d x 6" long
8d: 8 pennyweight x 1¾" long	spike = any nail 20d or larger
10d: 10 pennyweight x 2" long	x 6" long

SURFACE FINISHING ABBREVIATIONS, YARD AND STRUCTURAL LUMBER

Abbreviation	Definition
S1S	Smooth surface, one side
S1S1E	Smooth surface, one side, one end
S1S2E	Smooth surface, one side, two ends
S2S	Smooth surface, two sides
S2S1E	Smooth surface, two sides, one end
S2S2E	Smooth surface, two sides, two ends
S3S	Smooth surface, three sides
S3S1E	Smooth surface, three sides, one end
S3S2E	Smooth surface, three sides, two ends
S4S	Smooth surface, four sides
S4S1E	Smooth surface, four sides, one end
S4S2E	Smooth surface, four sides, two ends
R/E	Resawn (same as S1S or S1S2E)
R/O or R/S	Rough sawn (no smooth surfaces)

GLUE-LAMINATED TIMBER SIZING

Width		Depth	
Nominal	Actual	Nominal	Actual
3"	2¼"	8" (6¾")	9"
4"	3⅛"	10" (5⅛")	10½"
6"	5⅛"	8" (5⅛")	9"
8"	6¾"	6" (5⅛")	6"
10"	8¾"	6" (3⅛")	7½"
12"	10¾"		
14"	12¼"		
16"	14¼"		

REBAR (Standard Sizes)

Bar	Diameter (size)	Bar	Diameter (size)
#2	¼"	#8	1"
#3	⅜"	#9	1⅛"
#4	½"	#10	1¼"
#5	⅝"	#11	1⅜"
#6	¾"	#14	1¾"
#7	⅞"	#18	2¼"

WELDED WIRE FABRIC

Roll	Sheet
6 x 6 – W1.4 x W1.4	6 x 6 – W2.0 x W2.0
6 x 6 – W2.9 x W2.9	6 x 6 – W2.9 x W2.9
6 x 6 – W4.0 x W4.0	6 x 6 – W4.0 x W4.0
6 x 6 – W5.5 x W5.5	4 x 4 – W1.4 x W1.4
4 x 4 – W1.4 x W1.4	4 x 4 – W2.9 x W2.9
4 x 4 – W2.9 x W2.9	
4 x 4 – W4.0 x W4.0	

If the letter designation D is added, the fabric is deformed similar to rebar. WWF is made into sheets or rolls depending upon the size and quantity required.

FRAMING RATIOS PER STUD SPACING

Spacing	Constant	Decimal Equivalent
12" O/C	1	1.000
16" O/C	¾	0.750
18" O/C	⅔	0.667
20" O/C	⅗	0.600
24" O/C	½	0.500
32" O/C	⅜	0.375
36" O/C	⅓	0.333

HARDWARE LOCATIONS

Locks, Latches, Roller Latches and Double Handle Sets	Centerline of lock strike 40⁵⁄₁₆" from bottom of frame
Cylindrical or Mortise Deadlocks	Centerline of strike 60" from bottom of frame
Push Plates	Centerline 45" from bottom of frame
Pull Plates	Centerline of grip 42" from bottom of frame
Combination Push Bar	Centerline 42" from bottom of frame
Hospital Arm Pull	Centerline of lower base is 45" from bottom of frame with grip open at bottom
Panic Devices	Centerline of strike 40⁵⁄₁₆" from bottom of frame
Top Hinge	Up to 11³⁄₄" from rabbet section of head of frame to centerline of hinge
Bottom Hinge	Up to 13" from bottom of frame to centerline of hinge
Intermediate Hinge(s)	Equally spaced between top and bottom hinge

HARDWARE REINFORCING GAUGES

Hardware		Minimum Gauge
Hinges	1¾" Door	10 gauge or equivalent number of threads
		12 gauge allowed if reinforcing is channel shaped
	1¾" Frame	10 gauge or equivalent number of threads
	1⅜" Door	12 gauge or equivalent number of threads
	1⅜" Frame	12 gauge or equivalent number of threads
Mortise Locksets and Deadlocks	Door	14 gauge or equivalent number of threads
	Frame	14 gauge or equivalent number of threads
Bored or Cylindrical Locks	Door	14 gauge or equivalent number of threads
	Frame	14 gauge or equivalent number of threads
Flush Bolts and Chain and Foot Bolts	Door	14 gauge
	Frame	14 gauge
Surface Applied Closers	Door	14 gauge
	Frame	14 gauge
Hold-Open Arms	Door	14 gauge
	Frame	14 gauge
Pull Plates and Bars Kick and Push Plates and Bars	Door	16 gauge except when through bolts are used
	Door	Not required
Surface Panic Devices	Door	14 gauge
	Frame	14 gauge
Floor Checking Hinges and Pivots	Door	7 gauge
	Frame	7 gauge

STEEL DOOR SELECTION GUIDE FOR INDUSTRIAL BUILDINGS

Industrial Bldgs.	Door Grades			Thickness		Door Design Nomenclature					General Remarks
	I	II	III	1¾	1⅜	F	G	V	FG	N	
Offices											
Entrance	•			•		•	•		•		Two-way vision
Individual Office	•			•		•	•				
Closet				•	•	•					
Bathroom		•	•	•		•					Option—louver
Stairwell		•	•	•				•			Fire door—two-way vision—check local codes
Manufacturing											
Entrance			•	•		•			•	•	Flush for security
Equipment Room		•	•	•		•					
Cafeteria		•	•	•			•				
Parts Crib			•	•		•					Option—Dutch door
Tool Room			•	•		•	•				Option—Dutch door
Boiler Room		•	•	•		•					Louver optional
Bathroom			•	•		•					Option—louvers

Door letter symbols: F=Flush G=Half Glass FG=Full Glass V=Vision Lite N=Narrow Lite

1-25

STEEL DOOR SELECTION GUIDE FOR INDUSTRIAL BUILDINGS (cont.)

Industrial Bldgs.	Door Grades			Thickness		Door Design Nomenclature					General Remarks
	I	II	III	1¾	1⅜	F	G	V	FG	N	
Entrance	•		•	•			•		•	•	Two-way vision
Individual Office	•	•	•	•	•	•	•				Optional thickness and design
Closet		•	•	•	•	•					
Toilet		•	•	•		•					Design option—louver door
Stairwell		•	•	•				•			Fire door—two-way vision—check local codes
Equipment Room		•	•	•		•					
Boiler Room		•	•	•		•					Louver optional
Door letter symbols: F=Flush G=Half Glass V=Vision Lite FG=Full Glass N=Narrow Lite											

STANDARD STEEL DOOR GRADES AND MODELS

Grade	Model	Construction	Minimum Thickness Gauge Number Panels – Face Sheet	Minimum Thickness Gauge Number Stiles – Rails
I – Standard Duty (1¾" and 1⅜")	1	Full Flush (Hollow Steel)	20	–
	2	Full Flush (Composite)	20	–
	3	Seamless (Hollow Steel)	20	–
	4	Seamless (Composite)	20	–
II – Heavy Duty (1¾" only)	1	Full Flush (Hollow Steel)	18	–
	2	Full Flush (Composite)	18	–
	3	Seamless (Hollow Steel)	18	–
	4	Seamless (Composite)	18	–
III – Extra Heavy Duty (1¾" only)	1	Full Flush (Hollow Steel)	16	–
	2	Full Flush (Composite)	16	–
	3	Seamless (Hollow Steel)	16	–
	4	Seamless (Composite)	16	–
	5	Flush Panel (Stile and Rail)	18	16

FIRE DOOR CLASSIFIED OPENINGS

Opening	Class	Rating		Glass
	A	3 Hours		None
	B	1½ Hours		100 sq. in. per door leaf
	C	¾ Hour		1296 sq. in. per light
	D	1½ Hours		None
	E	¾ Hour		1296 sq. in. per light
	No class designation	20 Minutes		1296 sq. in. per light

ROOF DEAD LOADS

Material	Weight, psf
Shingles:	
Asphalt (¼" approx.)	2.0
Book tile (2")	12.0
Book tile (3")	20.0
Cement asbestos (⅜" approx.)	4.0
Clay tile (for mortar add 10 psf)	9.0 to 14.0
Ludowici	10.0
Roman	12.0
Slate (¼")	10.0
Spanish	19.0
Wood (1")	3.0

CEILING DEAD LOADS

Material	Weight, psf
Acoustical fiber tile	1.0
Channel-suspended system	1.0

FLOOR DEAD LOADS

Material	Weight, psf
Hardwood (1" nominal)	4.0
Plywood (per inch of thickness)	3.0
Asphalt mastic (per inch of thickness)	12.0
Cement finish (per inch of thickness)	12.0
Ceramic and quarry tile (¾")	10.0
Concrete (per inch of thickness)	
Lightweight	6.0 to 10.0
Reinforced (normal weight)	12.5
Stone	12.0
Cork tile (1/16")	0.5
Flexicore (6" slab)	46.0
Linoleum (¼")	1.0
Terrazzo finish (1½")	19.0
Vinyl tile (⅛")	1.4

WALL AND PARTITION DEAD LOADS

Material	Weight, psf
Wood paneling (1")	2.5
Wood studs (2 x 4 DF-Larch)	
12" o.c.	1.3
16" o.c.	1.0
24" o.c.	0.7
Glass block (4")	18.0
Glass (¼" plate)	3.3
Glazed tile	18.0
Marble or marble wainscoting	15.0
Masonry (per 4" of thickness)	
Brick	38.0
Concrete block	30.0
Cinder concrete block	20.0
Hollow clay tile, load bearing	23.0
Hollow clay tile, non-load bearing	18.0
Hollow gypsum block	13.0
Limestone	55.0
Terra-cotta tile	25.0
Stone	55.0
Plaster (1")	8.0
Plaster (1") on wood lath	10.0
Plaster (1") on metal lath	8.5
Gypsum wallboard (1")	5.0
Porcelain-enameled steel	3.0
Stucco (⅞")	10.0
Windows (glass, frame, and sash)	8.0

AVERAGE WEIGHT OF HOUSE BY AREA

Unit	Load/ft.²
Roof	40 lb.
Attic (low)	20 lb.
Attic (full)	30 lb.
Second floor	30 lb.
First floor	40 lb.
Wall	12 lb.

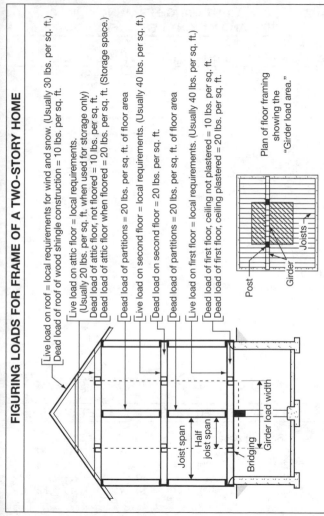

FIGURING LOADS FOR FRAME OF A TWO-STORY HOME

Live load on roof = local requirements for wind and snow. (Usually 30 lbs. per sq. ft.)
Dead load of roof of wood shingle construction = 10 lbs. per sq. ft.

Live load on attic floor = local requirements.
(Usually 20 lbs. per sq. ft. when used for storage only)
Dead load of attic floor, not floored = 10 lbs. per sq. ft.
Dead load of attic floor when floored = 20 lbs. per sq. ft. (Storage space.)

Dead load of partitions = 20 lbs. per sq. ft. of floor area
Live load on second floor = local requirements. (Usually 40 lbs. per sq. ft.)
Dead load on second floor = 20 lbs. per sq. ft.

Dead load of partitions = 20 lbs. per sq. ft. of floor area
Live load on first floor = local requirements. (Usually 40 lbs. per sq. ft.)
Dead load of first floor, ceiling not plastered = 10 lbs. per sq. ft.
Dead load of first floor, ceiling plastered = 20 lbs. per sq. ft.

Plan of floor framing
showing the
"Girder load area."

Joists

Girder

Post

Joist span

Half
joist span

Bridging

Girder load width

1-31

LIGHT GAUGE FRAMING, SIZE AND WEIGHT PER LINEAL FOOT

Studs

Size (inches)	Metal Gauge	Net Weight (lb./lf.)
1½"	25	0.443
1½"	20	0.700
2½"	25	0.509
2½"	20	0.810
3"	25	0.555
3"	20	0.875
3¼"	25	0.575
3¼"	20	0.910
3½"	25	0.597
3½"	20	0.944
3⅝"	25	0.608
3⅝"	20	0.964
4"	25	0.641
4"	20	1.014
5½"	20	1.240
6"	20	1.290

Joists

Size (inches)	Metal Gauge	Net Weight (lb./lf.)
6"	20	1.486
6"	18	1.889
6"	16	2.339
6"	14	2.914
8"	18	2.215
8"	16	2.411
10"	16	3.142
10"	14	3.908

RECOMMENDED TELECOM CLOSET SIZES

Serving Area		Closet Size	
Ft²	(m²)	ft. x ft.	(m x m)
10,000	(1000)	10 x 11	(3 x 3.4)
8,000	(800)	10 x 9	(3 x 2.8)
5,000	(500)	10 x 7	(3 x 2.2)

TREATED STRESSED SKIN PANEL CONSTRUCTION

Built-up roofing

Vapor barrier

Treated plywood stressed skin panels
Fire-retardant-treated plywood top skin
$\frac{3}{4}$" minimum thickness glued to
fire-retardant-treated joists

Blocking under plywood
joints unless scarfed

Bottom skin (optional) –
Fire-retardant-treated plywood
or gypsum board may be used

Untreated wood beams at least 8' 0" o.c.
(Trusses with heavy wood members permitted
in most states at this spacing).
Noncombustible supports may also be used.

Notes:
1. Aluminum foil vapor barrier required only for NM 501 construction.
2. For NM 501 construction, use tongue and groove plywood joints or treated blocking.

ONE-HOUR ASSEMBLY RESILIENT CHANNEL CEILING SYSTEM

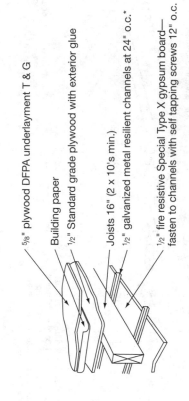

5/8" plywood DFPA underlayment T & G

Building paper

1/2" Standard grade plywood with exterior glue

Joists 16" (2 x 10's min.)

1/2" galvanized metal resilient channels at 24" o.c.*

1/2" fire resistive Special Type X gypsum board—
fasten to channels with self tapping screws 12" o.c.

*Channels may be suspended below joists.

1-34

ONE-HOUR ASSEMBLY T-BAR GRID CEILING SYSTEM

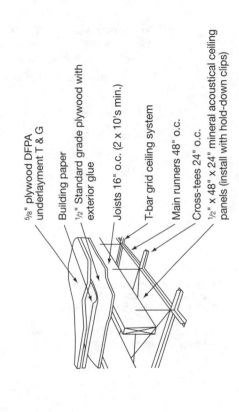

5⁄8" plywood DFPA underlayment T & G

Building paper

1⁄2" Standard grade plywood with exterior glue

Joists 16" o.c. (2 x 10's min.)

T-bar grid ceiling system

Main runners 48" o.c.

Cross-tees 24" o.c.

1⁄2" x 48" x 24" mineral acoustical ceiling panels (install with hold-down clips)

ONE-HOUR INTERIOR SHEAR
WALL CONSTRUCTION

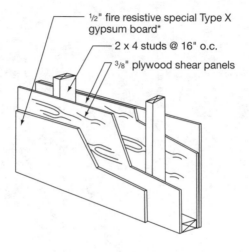

½" fire resistive special Type X
gypsum board*

2 x 4 studs @ 16" o.c.

³/₈" plywood shear panels

*Regular ½" gypsum board may be
used when mineral wool or glass
fiber batts are used in wall cavity.

Insulation batts in wall cavity also used
for sound transmission control.

ONE-HOUR EXTERIOR
WALL CONSTRUCTION

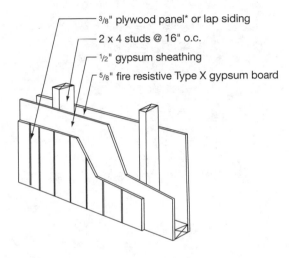

³/₈" plywood panel* or lap siding

2 x 4 studs @ 16" o.c.

¹/₂" gypsum sheathing

⁵/₈" fire resistive Type X gypsum board

*Including nominal ³/₈"
specialty plywood sidings

JOINT DETAILS FOR PLYWOOD SIDING

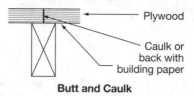

Plywood

Caulk or back with building paper

Butt and Caulk

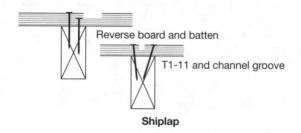

Reverse board and batten

T1-11 and channel groove

Shiplap

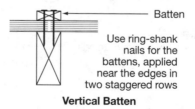

Batten

Use ring-shank nails for the battens, applied near the edges in two staggered rows

Vertical Batten

Vertical Wall Joints

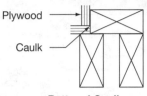

Plywood

Caulk

Butt and Caulk

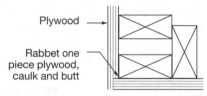

Plywood

Rabbet one
piece plywood,
caulk and butt

Rabbet and Caulk

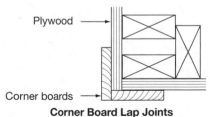

Plywood

Corner boards

Corner Board Lap Joints

Vertical Inside and Outside Corner Joints

JOINT DETAILS FOR PLYWOOD SIDING *(cont.)*

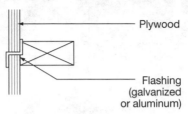

Plywood

Flashing
(galvanized
or aluminum)

Butt and Flash

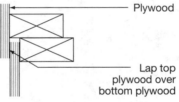

Plywood

Lap top
plywood over
bottom plywood

Lap Plywood

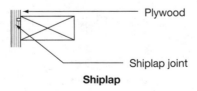

Plywood

Shiplap joint

Shiplap

Horizontal Wall Joints

JOINT DETAILS FOR PLYWOOD SIDING *(cont.)*

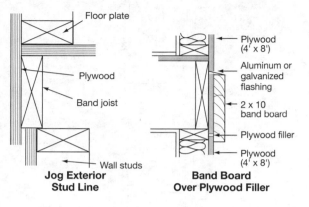

Jog Exterior Stud Line

Floor plate

Plywood

Band joist

Wall studs

Plywood (4' x 8')

Aluminum or galvanized flashing

2 x 10 band board

Plywood filler

Plywood (4' x 8')

Band Board Over Plywood Filler

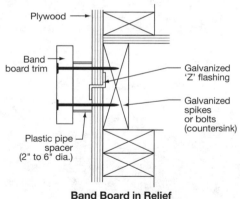

Plywood

Band board trim

Plastic pipe spacer (2" to 6" dia.)

Galvanized 'Z' flashing

Galvanized spikes or bolts (countersink)

Band Board in Relief

Horizontal Beltline Joints

Note: For multi-story buildings, make provisions at horizontal joints for settling shrinkage of framing, especially when applying siding direct to studs.

JOINT DETAILS FOR PLYWOOD SIDING *(cont.)*

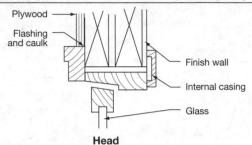

Plywood

Flashing
and caulk

Finish wall

Internal casing

Glass

Head

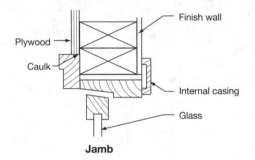

Finish wall

Plywood

Caulk

Internal casing

Glass

Jamb

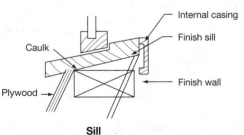

Internal casing

Finish sill

Caulk

Finish wall

Plywood

Sill

Window Details

LOADING TABLES FOR FLOOR/CEILING TRUSSES BUILT WITH 2 x 4 CHORDS

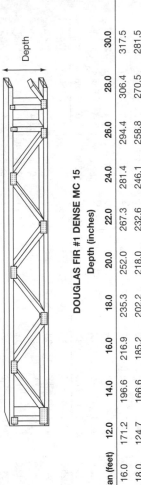

Depth

DOUGLAS FIR #1 DENSE MC 15

Depth (inches)

Maximum load*

Span (feet)	12.0	14.0	16.0	18.0	20.0	22.0	24.0	26.0	28.0	30.0
16.0	171.2	196.6	216.9	235.3	252.0	267.3	281.4	294.4	306.4	317.5
18.0	124.7	166.6	185.2	202.2	218.0	232.6	246.1	258.8	270.5	281.5
20.0	95.0	128.4	159.1	174.8	189.4	203.1	215.9	227.9	239.3	249.9
22.0	75.1	100.2	129.6	150.4	165.5	178.1	190.1	201.4	212.2	222.3
24.0	61.3	80.6	103.3	126.9	141.9	156.8	168.1	178.7	188.7	198.4
26.0	51.4	66.6	84.4	104.9	121.4	134.2	146.9	159.1	168.5	177.5
28.0	44.2	56.3	70.6	87.0	105.0	116.1	127.2	138.2	149.2	159.5
30.0	38.7	48.6	60.2	73.5	88.6	101.4	111.1	120.8	130.4	140.0
32.0	34.5	42.7	52.2	63.2	75.6	89.3	97.9	106.4	114.9	123.4
34.0	31.3	38.1	46.0	55.2	65.5	77.1	86.9	94.5	102.1	109.6

*Loads are shown in pounds per lineal foot. Deflection may govern the span limits.
These tables are limited to a simple span condition uniformly loaded.

1-43

LOADING TABLES FOR FLOOR/CEILING TRUSSES BUILT WITH 2 x 4 CHORDS *(cont.)*

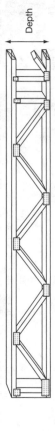

Depth

SOUTHERN PINE #1 DENSE KD

Maximum load*

Depth (inches)

Span (feet)	12.0	14.0	16.0	18.0	20.0	22.0	24.0	26.0	28.0	30.0
16.0	164.2	193.3	212.8	230.5	246.5	261.1	274.4	286.7	298.0	308.5
18.0	119.8	163.5	182.2	198.7	213.9	227.9	240.9	252.9	264.1	274.6
20.0	91.4	123.2	153.9	172.2	186.4	199.6	211.9	223.4	234.3	244.5
22.0	72.4	96.3	124.4	145.0	162.0	175.5	187.0	198.0	208.3	218.1
24.0	59.2	77.6	99.3	122.4	136.8	151.2	165.5	176.0	185.7	195.0
26.0	49.8	64.3	81.3	100.8	117.0	129.4	141.6	153.9	166.0	174.9
28.0	42.8	54.4	68.1	83.7	101.2	111.9	122.6	133.2	143.8	154.3
30.0	37.6	47.1	58.2	70.9	85.2	97.8	107.1	116.4	125.7	134.9
32.0	33.6	41.4	50.6	61.0	72.9	86.1	94.4	102.6	110.8	119.0
34.0	30.5	37.0	44.6	53.4	63.3	74.3	83.8	91.1	98.4	105.7

*Loads are shown in pounds per lineal foot. Deflection may govern the span limits.
These tables are limited to a simple span condition uniformly loaded.

LOADING TABLES FOR FLOOR/CEILING TRUSSES BUILT WITH 2 x 4 CHORDS *(cont.)*

Depth

MACHINE RATED LUMBER 2100F-1.8E

Maximum load*

Depth (inches)

Span (feet)	12.0	14.0	16.0	18.0	20.0	22.0	24.0	26.0	28.0	30.0
16.0	156.3	192.0	211.2	228.6	244.3	258.6	271.7	283.8	294.9	305.1
18.0	114.2	155.7	181.1	197.3	212.3	226.0	238.8	250.6	261.6	271.9
20.0	87.4	117.5	153.0	171.2	185.1	198.1	210.3	221.7	232.3	242.3
22.0	69.4	92.0	118.7	149.2	162.2	174.3	185.8	196.5	206.7	216.4
24.0	56.9	74.3	94.8	118.4	142.8	154.1	164.7	174.9	184.5	193.6
26.0	47.9	61.7	77.8	96.3	117.2	136.8	146.7	156.1	165.2	173.8
28.0	41.4	52.4	65.3	80.1	96.9	115.5	131.2	139.9	148.4	156.5
30.0	36.4	45.4	55.9	67.9	81.5	96.7	113.4	125.9	133.8	141.3
32.0	32.7	40.0	48.7	58.6	69.8	82.3	96.1	111.2	121.0	128.1
34.0	29.7	35.9	43.1	51.4	60.7	71.1	82.6	95.2	108.8	116.5

*Loads are shown in pounds per lineal foot. Deflection may govern the span limits. These tables are limited to a simple span condition uniformly loaded.

SHEET METAL SCREW CHARACTERISTICS

Screw Size #	Screw Diameter (inches)	Diameter of Pierced Hole (inches)	Hole Size	Thickness of Metal – Gauge #
#4	.112	.086	#44	28
		.086	#44	26
		.093	#42	24
		.098	#42	22
		.100	#40	20
#6	.138	.111	#39	28
		.111	#39	26
		.111	#39	24
		.111	#38	22
		.111	#36	20
#7	.155	.121	#37	28
		.121	#37	26
		.121	#35	24
		.121	#33	22
		.121	#32	20
		–	#31	18
#8	.165	.137	#33	26
		.137	#33	24
		.137	#32	22
		.137	#31	20
		–	#30	18
#10	.191	.158	#30	26
		.158	#30	24
		.158	#30	22
		.158	#29	20
		.158	#25	18
#12	.218	–	#26	24
		.185	#25	22
		.185	#24	20
		.185	#22	18
#14	.251	–	#15	24
		.212	#12	22
		.212	#11	20
		.212	#9	18

STANDARD WOOD SCREW CHARACTERISTICS (Inches)

Screw Size #	Wood Screw Standard Lengths	Size of Pilot Hole		Size of Shank Hole	
		Softwood Bit #	Hardwood Bit #	Clearance Bit #	Hole Diameter
0	¼"	75	66	52	.060"
1	¼" to ⅜"	71	57	47	.073"
2	¼" to ½"	65	54	42	.086"
3	¼" to ⅝"	58	53	37	.099"
4	⅜" to ¾"	55	51	32	.112"
5	⅜" to ¾"	53	47	30	.125"
6	⅜" to 1½"	52	44	27	.138"
7	⅜" to 1½"	51	39	22	.151"
8	½" to 2"	48	35	18	.164"
9	⅝" to 2¼"	45	33	14	.177"
10	⅝" to 2½"	43	31	10	.190"
11	¾" to 3"	40	29	4	.203"
12	⅞" to 3½"	38	25	2	.216"
14	1" to 4½"	32	14	D	.242"
16	1¼" to 5½"	29	10	I	.268"
18	1½" to 6"	26	6	N	.294"
20	1¾" to 6"	19	3	P	.320"
24	3½" to 6"	15	D	V	.372"

ALLEN HEAD AND MACHINE SCREW BOLT AND TORQUE CHARACTERISTICS

Number of Threads Per Inch	Allen Head and Mach. Screw Bolt Size	Allen Head Case H Steel 160,000 psi	Mach. Screw Yellow Brass 60,000 psi	Mach. Screw Silicone Bronze 70,000 psi
		Torque in Foot-Pounds or Inch-Pounds		
4.5	2"	8800	–	–
5	1¾"	6100	–	–
6	1½"	3450	655	595
6	1⅜"	2850	–	–
7	1¼"	2130	450	400
7	1⅛"	1520	365	325
8	1"	970	250	215
9	⅞"	640	180	160
10	¾"	400	117	104
11	⅝"	250	88	78
12	⁹⁄₁₆"	180	53	49
13	½"	125	41	37
14	⁷⁄₁₆"	84	30	27
16	⅜"	54	20	17
18	⁵⁄₁₆"	33	125 in#	110 in#
20	¼"	16	70 in#	65 in#
24	#10	60	22 in#	20 in#
32	#8	46	19 in#	16 in#
32	#6	21	10 in#	8 in#
40	#5	–	7.2 in#	6.4 in#
40	#4	–	4.9 in#	4.4 in#
48	#3	–	3.7 in#	3.3 in#
56	#2	–	2.3 in#	2 in#

For fine thread bolts, increase by 9%.

HEX HEAD BOLT AND TORQUE CHARACTERISTICS

BOLT MAKE-UP IS STEEL WITH COARSE THREADS

Number of Threads Per Inch	Hex Head Bolt Size	⬡ SAE 0-1-2 74,000 psi	⬡ SAE Grade 3 100,000 psi	⬡ SAE Grade 5 120,000 psi
		Torque = Foot-Pounds		
4.5	2"	2750	5427	4550
5	1¾"	1900	3436	3150
6	1½"	1100	1943	1775
6	1⅜"	900	1624	1500
7	1¼"	675	1211	1105
7	1⅛"	480	872	794
8	1"	310	551	587
9	⅞"	206	372	382
10	¾"	155	234	257
11	⅝"	96	145	154
12	⁹⁄₁₆"	69	103	114
13	½"	47	69	78
14	⁷⁄₁₆"	32	47	54
16	⅜"	20	30	33
18	⁵⁄₁₆"	12	17	19
20	¼"	6	9	10

For fine thread bolts, increase by 9%.

BOLT MAKE-UP IS STEEL WITH COARSE THREADS

Number of Threads Per Inch	Hex Head Bolt Size	SAE Grade 6 133,000 psi	SAE Grade 7 133,000 psi	SAE Grade 8 150,000 psi
		Torque = Foot-Pounds		
4.5	2"	7491	7500	8200
5	1¾"	5189	5300	5650
6	1½"	2913	3000	3200
6	1⅜"	2434	2500	2650
7	1¼"	1815	1825	1975
7	1⅛"	1304	1325	1430
8	1"	825	840	700
9	⅞"	550	570	600
10	¾"	350	360	380
11	⅝"	209	215	230
12	⁹⁄₁₆"	150	154	169
13	½"	106	110	119
14	⁷⁄₁₆"	69	71	78
16	⅜"	43	44	47
18	⁵⁄₁₆"	24	25	29
20	¼"	12.5	13	14

For fine thread bolts, increase by 9%.
For special alloy bolts, obtain torque rating from the manufacturer.

WHITWORTH HEX HEAD BOLT AND TORQUE CHARACTERISTICS

BOLT MAKE-UP IS STEEL WITH COARSE THREADS

Number of Threads Per Inch	Whitworth Type Hex Head Bolt Size	Grades A & B 62,720 psi	Grade S 112,000 psi	Grade T 123,200 psi	Grade V 145,600 psi
			Torque = Foot-Pounds		
8	1"	276	497	611	693
9	7/8"	186	322	407	459
11	3/4"	118	213	259	287
11	5/8"	73	128	155	175
12	9/16"	52	94	111	128
12	1/2"	36	64	79	89
14	7/16"	24	43	51	58
16	3/8"	15	27	31	36
18	5/16"	9	15	18	21
20	1/4"	5	7	9	10

For fine thread bolts, increase by 9%.

1-51

METRIC HEX HEAD BOLT AND TORQUE CHARACTERISTICS

BOLT MAKE-UP IS STEEL WITH COARSE THREADS

(Metric Type) Thread Pitch	(Dimensions in Millimeters) Bolt Size	Standard 5D 71,160 psi	Standard 8G 113,800 psi	Standard 10K 142,000 psi	Standard 12K 170,674 psi
		Torque = Foot-Pounds			
3.0	24	261	419	570	689
2.5	22	182	284	394	464
2.0	18	111	182	236	183
2.0	16	83	132	175	208
1.25	14	55	89	117	137
1.25	12	34	54	70	86
1.25	10	19	31	40	49
1.0	8	10	16	22	27
1.0	6	5	6	8	10

For fine thread bolts, increase by 9%.

TIGHTENING TORQUE IN POUND-FEET-SCREW FIT			
Wire Size, AWG/kcmil	Driver	Bolt	Other
18-16	1.67	6.25	4.2
14-8	1.67	6.25	6.125
6-4	3.0	12.5	8.0
3-1	3.2	21.00	10.40
0-2/0	4.22	29	12.5
3/0-200	–	37.5	17.0
250-300	–	50.0	21.0
400	–	62.5	21.0
500	–	62.5	25.0
600-750	–	75.0	25.0
800-1000	–	83.25	33.0
1250-2000	–	83.26	42.0

SCREW TORQUES	
Screw Size, Inches Across, Hex Flats	Torque, Pound-Feet
$\frac{1}{8}$	4.2
$\frac{5}{32}$	8.3
$\frac{3}{16}$	15
$\frac{7}{32}$	23.25
$\frac{1}{4}$	42

STANDARD TAPS AND DIES (In Inches)

Thread Size	Coarse			Fine		
	Drill Size	Threads Per Inch	Decimal Size	Drill Size	Threads Per Inch	Decimal Size
4"	3	4	3.75			
3¾"	3	4	3.5			
3½"	3	4	3.25			
3¼"	3	4	3.0			
3"	2	4	2.75			
2¾"	2	4	2.5			
2½"	2	4	2.25			
2¼"	2	4.5	2.0313			
2"	1	4.5	1.7813			
1¾"	1	2	1.5469			
1½"	1	6	1.3281	1²⁷⁄₆₄"	12	1.4219
1⅜"	1	6	1.2188	1¹⁹⁄₆₄"	12	1.2969
1¼"	1	7	1.1094	1¹¹⁄₆₄"	12	1.1719
1⅛"	⁶³⁄₆₄"	7	.9844	1³⁄₆₄"	12	1.0469
1"	⅞"	8	.8750	¹⁵⁄₁₆"	14	.9375
⅞"	⁴⁹⁄₆₄"	9	.7656	¹³⁄₁₆"	14	.8125
¾"	²¹⁄₃₂"	10	.6563	¹¹⁄₁₆"	16	.6875
⅝"	¹⁷⁄₃₂"	11	.5313	³⁷⁄₆₄"	18	.5781
⁹⁄₁₆"	³¹⁄₆₄"	12	.4844	³³⁄₆₄"	18	.5156
½"	²⁷⁄₆₄"	13	.4219	²⁹⁄₆₄"	20	.4531
⁷⁄₁₆"	U	14	.368	²⁵⁄₆₄"	20	.3906
⅜"	⁵⁄₁₆"	16	.3125	Q	24	.332
⁵⁄₁₆"	F	18	.2570	I	24	.272
¼"	#7	20	.201	#3	28	.213
#12	#16	24	.177	#14	28	.182
#10	#25	24	.1495	#21	32	.159
³⁄₁₆"	#26	24	.147	#22	32	.157
#8	#29	32	.136	#29	36	.136
#6	#36	32	.1065	#33	40	.113
#5	#38	40	.1015	#37	44	.104
⅛"	³⁄₃₂"	32	.0938	#38	40	.1015
#4	#43	40	.089	#42	48	.0935
#3	#47	48	.0785	#45	56	.082
#2	#50	56	.07	#50	64	.07
#1	#53	64	.0595	#53	72	.0595
#0	–	–	–	³⁄₆₄"	80	.0469

(mm) Thread Pitch	TAPS AND DIES – METRIC CONVERSIONS			
	Fine Thread Size		Tap Drill Size	
	Inches	mm	Inches	mm
4.5	1.6535	42	1.4567	37.0
4.0	1.5748	40	1.4173	36.0
4.0	1.5354	39	1.3779	35.0
4.0	1.4961	38	1.3386	34.0
4.0	1.4173	36	1.2598	32.0
3.5	1.3386	34	1.2008	30.5
3.5	1.2992	33	1.1614	29.5
3.5	1.2598	32	1.1220	28.5
3.5	1.1811	30	1.0433	26.5
3.0	1.1024	28	.9842	25.0
3.0	1.0630	27	.9449	24.0
3.0	1.0236	26	.9055	23.0
3.0	.9449	24	.8268	21.0
2.5	.8771	22	.7677	19.5
2.5	.7974	20	.6890	17.5
2.5	.7087	18	.6102	15.5
2.0	.6299	16	.5118	14.0
2.0	.5512	14	.4724	12.0
1.75	.4624	12	.4134	10.5
1.50	.4624	12	.4134	10.5
1.50	.3937	11	.3780	9.6
1.50	.3937	10	.3386	8.6
1.25	.3543	9	.3071	7.8
1.25	.3150	8	.2677	6.8
1.0	.2856	7	.2362	6.0
1.0	.2362	6	.1968	5.0
.90	.2165	5.5	.1811	4.6
.80	.1968	5	.1653	4.2
.75	.1772	4.5	.1476	3.75
.70	.1575	4	.1299	3.3
.75	.1575	4	.1279	3.25
.60	.1378	3.5	.1142	2.9
.60	.1181	3	.0945	2.4
.50	.1181	3	.0984	2.5
.45	.1124	2.6	.0827	2.1
.45	.0984	2.5	.0787	2.0
.40	.0895	2.3	.0748	1.9
.40	.0787	2	.0630	1.6
.45	.0787	2	.0590	1.5
.35	.0590	1.5	.0433	1.1

RECOMMENDED DRILLING SPEEDS (RPMS)

Material	Bit Sizes	RPM Speed Range		
Glass	Special Metal Tube Drilling	700		
Plastics	7/16" and larger 3/8" 5/16" 1/4" 3/16" 1/8" 1/16" and smaller	500 1500 2000 3000 3500 5000 6000	– – – – – – –	1000 2000 2500 3500 4000 6000 6500
Woods	1" and larger 3/4" to 1" 1/2" to 3/4" 1/4" to 1/2" 1/4" and smaller carving / routing	700 2000 2300 3100 3800 4000	– – – – – –	2000 2300 3100 3800 4000 6000
Soft Metals	7/16" and larger 3/8" 5/16" 1/4" 3/16" 1/8" 1/16" and smaller	1500 3000 3500 4500 5000 6000 6000	– – – – – – –	2500 3500 4000 5000 6000 6500 6500
Steel	7/16" and larger 3/8" 5/16" 1/4" 3/16" 1/8" 1/16" and smaller	500 1000 1000 1500 2000 3000 5000	– – – – – – –	1000 1500 1500 2000 2500 4000 6500
Cast Iron	7/16" and larger 3/8" 5/16" 1/4" 3/16" 1/8" 1/16" and smaller	1000 1500 1500 2000 2500 3500 6000	– – – – – – –	1500 2000 2000 2500 3000 4500 6500

TORQUE LUBRICATION EFFECTS IN FOOT-POUNDS

Lubricant	$5/16"$ – 18 Thread	$1/2"$ – 13 Thread	Torque Decrease
Graphite	13	62	49 – 55%
Mily Film	14	66	45 – 52%
White Grease	16	79	35 – 45%
SAE 30	16	79	35 – 45%
SAE 40	17	83	31 – 41%
SAE 20	18	87	28 – 38%
Plated	19	90	26 – 34%
No Lube	29	121	0%

METALWORKING LUBRICANTS

Materials	Threading	Lathing	Drilling
Machine Steels	Dissolvable Oil Mineral Oil Lard Oil	Dissolvable Oil	Dissolvable Oil Sulpherized Oil Min. Lard Oil
Tool Steels	Lard Oil Sulpherized Oil	Dissolvable Oil	Dissolvable Oil Sulpherized Oil
Cast Irons	Sulpherized Oil Dry Min. Lard Oil	Dissolvable Oil Dry	Dissolvable Oil Dry Air Jet
Malleable Irons	Soda Water Lard Oil	Soda Water Dissolvable Oil	Soda Water Dry
Aluminums	Kerosene Dissolvable Oil Lard Oil	Dissolvable Oil	Kerosene Dissolvable Oil
Brasses	Dissolvable Oil Lard Oil	Dissolvable Oil	Kerosene Dissolvable Oil Dry
Bronzes	Dissolvable Oil Lard Oil	Dissolvable Oil	Dissolvable Oil Dry
Coppers	Dissolvable Oil Lard Oil	Dissolvable Oil	Kerosene Dissolvable Oil Dry

STANDARD HAND SIGNALS FOR CONTROLLING CRANE OPERATIONS

HOIST. Forearm vertical, forefinger pointing up, move hand in small horizontal circles.

LOWER. Arm extended downward, forefinger pointing down, move hand in small horizontal circles.

USE MAIN HOIST. Tap fist on head; then use regular signals.

USE WHIPLINE. Tap elbow with one hand; then use regular signals.

RAISE BOOM. Arm extended, fingers closed, thumb pointing upward.

LOWER BOOM. Arm extended, fingers closed, thumb pointing down.

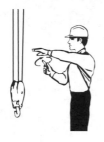

MOVE SLOWLY. One hand gives motion signal, other hand motionless in front of hand giving the motion signal.

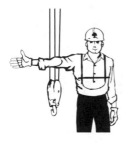

RAISE THE BOOM AND LOWER THE LOAD. Arm extended, thumb pointing up, flex fingers in and out.

LOWER THE BOOM AND RAISE THE LOAD. Arm extended, thumb pointing down, flex fingers in and out.

SWING. Arm extended, point with finger in direction of swing.

STOP. Arm extended, palm down, hold.

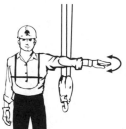

EMERGENCY STOP. Arm extended, palm down, move hand rapidly right and left.

STANDARD HAND SIGNALS FOR
CONTROLLING CRANE OPERATIONS *(cont.)*

TRAVEL. Arm extended forward, hand open and slightly raised, pushing motion in direction of travel.

EXTEND BOOM. Both fists in front of body with thumbs pointing outward.

RETRACT BOOM. Both fists in front of body with thumbs pointing toward each other.

MANILA ROPE SLINGS – STRAIGHT LEG

Rope Diameter Inches	Straight 2 Leg	Straight 4 Leg	Choker 2 Leg	60° Choker 4 Leg	45° Choker 4 Leg	30° Choker 4 Leg
			TONS CAPACITY			
1/2"	1/2	1	1/3	2/3	1/2	1/3
3/4"	3/4	1 1/2	3/4	1 1/4	1	3/4
1"	1 1/2	3	1 1/4	2	1 1/2	1 1/4
1 1/2"	3	6	2	4	3	2
2"	5	10	4	7	6	4
2 1/2"	7	15	6	10	8	6
3"	10	20	8	14	12	8
3 1/2"	14	29	11	20	16	11
4"	17	34	13	23	19	13

MANILA ROPE SLINGS – BASKET

Rope Diameter Inches	60° Basket 4 Leg	45° Basket 4 Leg	30° Basket 4 Leg	60° Basket 6 Leg	45° Basket 6 Leg	30° Basket 6 Leg
			TONS CAPACITY			
$\frac{1}{2}$"	$\frac{3}{4}$	$\frac{2}{3}$	$\frac{1}{2}$	1	$\frac{3}{4}$	$\frac{2}{3}$
$\frac{3}{4}$"	$1\frac{1}{2}$	1	$\frac{3}{4}$	$2\frac{1}{4}$	2	1
1"	$2\frac{1}{2}$	2	$1\frac{1}{2}$	4	3	2
$1\frac{1}{2}$"	5	4	3	8	6	4
2"	9	7	5	13	11	7
$2\frac{1}{2}$"	13	11	7	19	16	11
3"	18	15	10	27	22	15
$3\frac{1}{2}$"	25	21	15	38	31	22
4"	30	24	17	44	36	25

WIRE ROPE SLINGS – STRAIGHT LEG

Rope Diameter Inches	Straight 1 Leg	Choker 1 Leg	60° Choker 2 Leg	45° Choker 2 Leg	30° Choker 2 Leg
			TONS CAPACITY		
1/4"	1/2	1/3	2/3	1/2	1/3
3/8"	1	3/4	1 1/4	1	3/4
1/2"	2	1 1/2	2 1/2	2	1 1/2
5/8"	3	2	4	3	2
3/4"	4	3	5	4	3
1	7	5	8	7	5
1 1/4"	10	7	12	9	7
1 1/2"	13	9	16	13	9
2"	21	15	27	22	15
2 1/2"	28	22	38	31	22
3"	36	28	49	40	28
3 1/2"	40	34	59	48	34

WIRE ROPE SLINGS – BASKET

Rope Diameter Inches	60° Basket 2 Leg	45° Basket 2 Leg	30° Basket 2 Leg	60° Basket 4 Leg	45° Basket 4 Leg	30° Basket 4 Leg
			TONS CAPACITY			
1/4"	2/3	1/2	1/3	1	1	3/4
3/8"	1 1/2	1	3/4	3	2	1 1/2
1/2"	2 1/2	2	1 1/2	5	4	3
5/8"	4	3	2	7	6	4
3/4"	5	4	3	11	9	6
1"	9	7	5	18	15	10
1 1/4"	13	11	7	26	21	15
1 1/2"	17	14	10	35	28	20
2"	27	22	15	53	44	31
2 1/2"	38	31	22	75	61	43
3"	49	40	29	97	80	56
3 1/2"	59	49	34	118	97	68

1-66

SAFE LOADS FOR SHACKLES

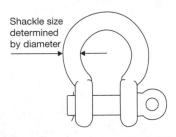

Shackle size determined by diameter

Size (inches)	Safe Load at 90° (tons)
$\frac{1}{4}$	$\frac{1}{3}$
$\frac{5}{16}$	$\frac{1}{2}$
$\frac{3}{8}$	$\frac{3}{4}$
$\frac{7}{16}$	1
$\frac{1}{2}$	$1\frac{1}{2}$
$\frac{5}{8}$	2
$\frac{3}{4}$	3
$\frac{7}{8}$	4
1	$5\frac{1}{2}$
$1\frac{1}{8}$	$6\frac{1}{2}$
$1\frac{1}{4}$	8
$1\frac{3}{8}$	10
$1\frac{1}{2}$	12
$1\frac{3}{4}$	16
2	21
$2\frac{1}{4}$	27
$2\frac{1}{2}$	34
$2\frac{3}{4}$	40
3	50

EYE AND FACE PROTECTORS

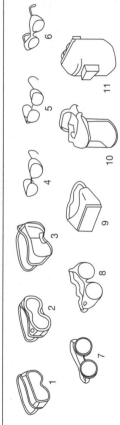

1. GOGGLES, Flexible Fitting – Regular Ventilation
2. GOGGLES, Flexible Fitting – Hooded Ventilation
3. GOGGLES, Cushioned Fitting – Rigid Body
4. SPECTACLES, Metal Frame – with Sideshields[1]
5. SPECTACLES, Plastic Frame – with Sideshields
6. SPECTACLES, Metal-Plastic Frame – with Sideshields[1]
7. WELDING GOGGLES, Eyecup Type – Tinted Lenses
7A. CHIPPING GOGGLES, Eyecup Type – Clear Safety Lenses
8. WELDING GOGGLES, Coverspec Type – Tinted Lenses
8A. CHIPPING GOGGLES, Coverspec Type – Clear Safety Lenses
9. WELDING GOGGLES, Coverspec Type – Tinted Plate Lens
10. FACE SHIELD (Available with Plastic or Mesh Window)
11. WELDING HELMETS

[1]Non-side shield spectacles are available for limited hazard use requiring only frontal protection.

EYE PROTECTION

Operation	Hazards	Recommended protectors
Acetylene-Burning, Acetylene-Cutting, Acetylene-Welding	Sparks, harmful rays, molten metal, flying particles	7, 8, 9
Chemical Handling	Splash, acid burns, fumes	2, 10 (For severe exposure add 10 over 2)
Chipping	Flying particles	1, 3, 4, 5, 6, 7A, 8A
Electric (arc) welding	Sparks, intense rays, molten metal	9, 11, (11 in combination with 4, 5, 6, in tinted lenses advisable)
Furnace operations	Glare, heat, molten metal	7, 8, 9 (For severe exposure add 10)
Grinding-Light	Flying particles	1, 3, 4, 5, 6, 10
Grinding-Heavy	Flying particles	1, 3, 7A, 8A (For severe exposure add 10)
Laboratory	Chemical splash, glass breakage	2 (10 when in combination with 4, 5, 6)
Machining	Flying particles	1, 3, 4, 5, 6, 10
Molten metals	Heat, glare, sparks, splash	7, 8, (10 in combination with 4, 5, 6, in tinted lenses)
Spot welding	Flying particles, sparks	1, 3, 4, 5, 6, 10

FILTER LENS SHADE NUMBERS FOR PROTECTION AGAINST RADIANT ENERGY

Welding operation	Shade number
Shielded metal-arc welding 1/16-, 3/32-, 1/8-, 5/32-inch diameter electrodes	10
Gas-shielded arc welding (nonferrous) 1/16-, 3/32-, 1/8-, 5/32-inch diameter electrodes	11
Gas-shielded arc welding (ferrous) 1/16-, 3/32-, 1/8-, 5/32-inch diameter electrodes	12
Shielded metal-arc welding 3/16-, 7/32-, 1/4-inch diameter electrodes	12
5/16-, 3/8-inch diameter electrodes	14
Atomic hydrogen welding	10 –14
Carbon-arc welding	14
Soldering	2
Torch brazing	3 or 4
Light cutting, up to 1 inch	3 or 4
Medium cutting, 1 inch to 6 inches	4 or 5
Heavy cutting, over 6 inches	5 or 6
Gas welding (light), up to 1/8-inch	4 or 5
Gas welding (medium), 1/8-inch to 1/2-inch	5 or 6
Gas welding (heavy), over 1/2-inch	6 or 8

SELECTION OF RESPIRATORS

Hazard	Respirator
Oxygen deficiency	Self-contained breathing apparatus. Hose mask with blower. Combination air-line respirator with auxiliary self-contained air supply or an air-storage receiver with alarm.
Gas and vapor contaminants immediately dangerous to life and health	Self-contained breathing apparatus. Hose mask with blower. Air-purifying full facepiece respirator (for escape only). Combination air-line respirator with auxiliary self-contained air supply or an air-storage receiver with alarm.
Not immediately dangerous to life and health	Air-line respirator. Hose mask without blower. Air-purifying, half-mask or mouthpiece respirator with chemical cartridge.
Particulate contaminants immediately dangerous to life and health	Self-contained breathing apparatus. Hose mask with blower. Air-purifying, full facepiece respirator with appropriate filter. Self-rescue mouthpiece respirator (for escape only). Combination air-line respirator with auxiliary self-contained air supply or an air-storage receiver with alarm.
Not immediately dangerous to life and health	Air-purifying, half-mask or mouthpiece respirator with filter pad or cartridge. Air-line respirator. Air-line abrasive-blasting respirator. Hose-mask without blower.

Immediately dangerous to life and health is a condition that either poses an immediate threat of severe exposure to contaminants such as radioactive materials, which are likely to have adverse delayed effects on health.

SELECTION OF RESPIRATORS *(cont.)*

Hazard	Respirator
Combination gas, vapor and particulate contaminants immediately dangerous to life and health	Self-contained breathing apparatus. Hose mask with blower. Air-purifying, full facepiece respirator with chemical canister and appropriate filter (gas mask with filter). Self-rescue mouthpiece respirator (for escape only). Combination air-line respirator with auxiliary self-contained air-supply or an air-storage receiver with alarm.
Not immediately dangerous to life and health	Air-line respirator. Hose mask without blower. Air-purifying, half-mask or mouthpiece respirator with chemical cartridge and appropriate filter.

Immediately dangerous to life and health is a condition that either poses an immediate threat of severe exposure to contaminants such as radioactive materials, which are likely to have adverse delayed effects on health.

PERMISSIBLE NOISE EXPOSURES

Duration per Day, Hours	Sound Level dBA Slow Response
8	90
6	92
4	95
3	97
2	100
1½	102
1	105
½	110
¼ or less	115

MINIMUM ILLUMINATION INTENSITIES IN FOOT-CANDLES

Foot-candles	Area of Operation
5	General construction area lighting.
3	General construction areas, concrete placement, excavation and waste areas, access ways, active storage areas, loading platforms, refueling, and field maintenance areas.
5	Indoors: warehouses, corridors, hallways, and exitways.
5	Tunnels, shafts, and general underground work areas. (Exception: minimum of 10 foot-candles is required at tunnel and shaft heading during drilling, mucking, and scaling. Bureau of Mines approved cap lights shall be acceptable for use in the tunnel heading.)
10	General construction plant and shops. (e.g., batch plants, screening plants, mechanical and electrical equipment rooms, carpenter shops, rigging lofts and active store rooms, mess halls, and indoor toilets and workrooms)
30	First aid stations, infirmaries, and offices.

FIRE-RESISTIVE LABELED EQUIPMENT

Product Classification	SMNA Spec.	SMNA Class	UL Equiv.	Product Design and Test Features
Fire-Insulated Safe	F 1-D	A	A	4 Hour Tested Fire-Resistive Safe (with impact test)
Fire-Insulated Safe	F 1-D	B	B	2 Hour Tested Fire-Resistive Safe (with impact test)
Fire-Insulated Safe	F 1-D	C	C	1 Hour Tested Fire-Resistive Safe (with impact test)
Fire-Insulated Record Container	F 1-D	C	C	1 Hour Tested Fire-Resistive Container (with impact test)
Fire-Insulated Safe	F 1-ND	D	D	1 Hour Tested Fire-Resistive Safe (without impact test)
Fire-Insulated Ledger Tray	F 1-D	C	C	1 Hour Tested Fire-Resistive Ledger Tray (with impact test)
Fire-Insulated Container	F 2-ND	E	E	½ Hour Tested Fire-Resistive Container (without impact test)
Fire-Insulated Container	F 2-ND	D	D	1 Hour Tested Fire-Resistive Container (without impact test)

Fire-Insulated Container	F 2-ND	2 Hour	2 Hour	B	2 Hour Tested Fire-Resistive Container (without impact test)
Fire-Insulated Vault Door	F 3	2 Hour	2 Hour	2 Hour	2 Hour Tested Fire-Resistive Vault Door
Fire-Insulated Vault Door	F 3	4 Hour	4 Hour	4 Hour	4 Hour Tested Fire-Resistive Vault Door
Fire-Insulated Vault Door	F 3	6 Hour	6 Hour	6 Hour	6 Hour Tested Fire-Resistive Vault Door
Fire-Insulated File Room Door	F 4	1 Hour	1 Hour	1 Hour	1 Hour Tested Fire-Resistive File or Storage Room Door
Fire-Insulated Record Container Data Processing Safe	F 2-D	Class 150	Class 150	Class 150	2 Hour or 4 Hour Fire-Resistive Data Processing Safe

Class A   Protects paper records from damage by fire (2,000° F) up to 4 hours.

Class B   Protects paper records from damage by fire (1,850° F) up to 2 hours.

Class C and D   Protects paper records from damage by fire (1,700° F) up to 1 hour.

Class E   Protects paper records from damage by fire (1,550° F) up to ½ hour.

Class 150   Protects EDP records from damage by fire and humidity for a rated period.

The Drop (or Impact) **Test:**   Used to determine whether or not the fire-resistance of a product would be impaired by being dropped 30 feet while still hot.

Fire-resistant equipment is designed specifically to resist fire, and consists of a metal shell filled with a fire-resistant insulation.

TYPES OF FIRE EXTINGUISHERS

Today they are virtually standard equipment in a business or residence and are rated by the makeup of the fire they will extinguish.

TYPE A: To extinguish fires involving trash, cloth, paper and other wood- or pulp-based materials. The flames are put out by water-based ingredients or dry chemicals.

TYPE B: To extinguish fires involving greases, paints, solvents, gas and other petroleum-based liquids. The flames are put out by cutting off oxygen and stopping the release of flammable vapors. Dry chemicals, foams and halon are used.

TYPE C: To extinguish fires involving electricity. The combustion is put out the same way as with a type B extinguisher but, most importantly, the chemical in a type C <u>MUST</u> be non-conductive to electricity in order to be safe and effective.

TYPE D: To extinguish fires involving combustible metals. Please be advised to obtain important information from your local fire department on the requirements for type D fire extinguishers for your area.

Any combination of letters indicate that an extinguisher will put out more than one type of fire. A type BC will put out two types of fires. The size of the fire to be extinguished is shown by a number in front of the letter such as 100A.
The following formulas apply:

Class 1A will extinguish 25 burning sticks 40 inches long.

Class 1B will extinguish a paint thinner fire 2.5 square feet in size.

A 100B fire extinguisher will put out a fire 100 times larger than a type 1B.

Here are some basic guidelines to follow:
• By using a type ABC you will cover most basic fires.
• Use fire extinguishers with a gauge and ones that are constructed with metal. Also note if the unit is U.L. approved.
• Utilize more than one extinguisher and be sure that each unit is mounted in a clearly visible and accessible manner.
• After purchasing any fire extinguisher always review the basic instructions for its intended use. Never deviate from the manufacturer's guidelines. Following this simple procedure could end up saving lives.

SIZE OF EXTENSION CORDS FOR PORTABLE TOOLS

Cord Length, Feet	Full-Load Rating of the Tool in Amperes at 115 Volts					
	0 to 2.0	2.1 to 3.4	3.5 to 5.0	5.1 to 7.0	7.1 to 12.0	12.1 to 16.0
	Wire Size (AWG)					
25	18	18	18	16	14	14
50	18	18	18	16	14	12
75	18	18	16	14	12	10
100	18	16	14	12	10	8
200	16	14	12	10	8	6
300	14	12	10	8	6	4
400	12	10	8	6	4	4
500	12	10	8	6	4	2
600	10	8	6	4	2	2
800	10	8	6	4	2	1
1000	8	6	4	2	1	0

SCAFFOLD MAXIMUM VERTICAL TIE SPACING WIDER THAN 3' 0" BASES

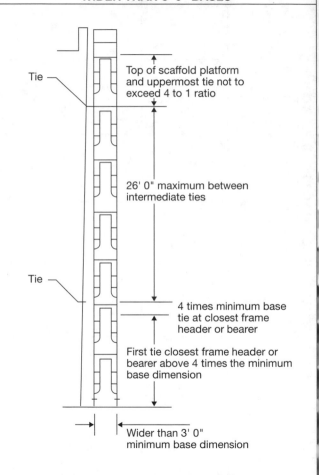

Tie

Top of scaffold platform and uppermost tie not to exceed 4 to 1 ratio

26' 0" maximum between intermediate ties

Tie

4 times minimum base tie at closest frame header or bearer

First tie closest frame header or bearer above 4 times the minimum base dimension

Wider than 3' 0" minimum base dimension

1-78

SINGLE POLE WOOD POLE SCAFFOLDS

	Light duty up to 20 feet high	Light duty up to 60 feet high	Medium duty up to 60 feet high	Heavy duty up to 60 feet high
Maximum intended load (lbs/ft^2)	25	25	50	75
Poles or uprights	2 x 4 in.	4 x 4 in.	4 x 4 in.	4 x 6 in.
Maximum pole spacing (longitudinal)	6 feet	10 feet	8 feet	6 feet
Maximum pole spacing (transverse)	5 feet	5 feet	5 feet	5 feet
Runners	1 x 4 in.	1¼ x 9 in.	2 x 10 in.	2 x 10 in.
Bearers and maximum spacing of bearers:				
3 feet	2 x 4 in.	2 x 4 in.	2 x 10 in. or 3 x 4 in.	2 x 10 in. or 3 x 5 in.
5 feet	2 x 6 in. or 3 x 4 in.	2 x 6 in. or 3 x 4 in. (rough)	2 x 10 in. or 3 x 4 in.	2 x 10 in. or 3 x 5 in.
6 feet			2 x 10 in. or 3 x 4 in.	2 x 10 in. or 3 x 5 in.
8 feet			2 x 10 in. or 3 x 4 in.	

SINGLE POLE WOOD POLE SCAFFOLDS *(cont.)*

	Light duty up to 20 feet high	Light duty up to 60 feet high	Medium duty up to 60 feet high	Heavy duty up to 60 feet high
Planking	1¼ x 9 in.	2 x 10 in.	2 x 10 in.	2 x 10 in.
Maximum vertical spacing of horizontal members	7 feet	9 feet	7 feet	6 ft. 6 in.
Bracing horizontal	1 x 4 in.	1 x 4 in.	1 x 6 in. or 1¼ x 4 in.	2 x 4 in.
Bracing diagonal	1 x 4 in.	1 x 4 in.	1 x 4 in.	2 x 4 in.
Tie-ins	1 x 4 in.	1 x 4 in.	1 x 4 in.	1 x 4 in.

Note: All members except planking are used on edge. All wood bearers shall be reinforced with $\frac{3}{16}$ x 2 inch steel strip, or the equivalent, secured to the lower edges for the entire length of the bearer.

INDEPENDENT WOOD POLE SCAFFOLDS

	Light duty up to 20 feet high	Light duty up to 60 feet high	Medium duty up to 60 feet high	Heavy duty up to 60 feet high
Maximum intended load (lbs/ft^2)	25	25	50	75
Poles or uprights	2 x 4 in.	4 x 4 in.	4 x 4 in.	4 x 4 in.
Maximum pole spacing (longitudinal)	6 feet	10 feet	8 feet	6 feet
Maximum pole spacing (transverse)	6 feet	10 feet	8 feet	8 feet
Runners	1¼ x 4 in.	1¼ x 9 in.	2 x 10 in.	2 x 10 in.
Bearers and maximum spacing of bearers:				
3 feet	2 x 4 in.	2 x 4 in.	2 x 10 in.	2 x 10 in. (rough)
6 feet	2 x 6 in. or 3 x 4 in.	2 x 10 in. (rough) or 3 x 8 in.	2 x 10 in.	2 x 10 in. (rough)
8 feet	2 x 6 in. or 3 x 4 in.	2 x 10 in. (rough) or 3 x 8 in.	2 x 10 in.	
10 feet	2 x 6 in. or 3 x 4 in.	2 x 10 in. (rough) or 3 x 8 in.		

INDEPENDENT WOOD POLE SCAFFOLDS (cont.)

	Light duty up to 20 feet high	Light duty up to 60 feet high	Medium duty up to 60 feet high	Heavy duty up to 60 feet high
Planking	1¼ x 9 in.	2 x 10 in.	2 x 10 in.	2 x 10 in.
Maximum vertical spacing of horizontal members	7 feet	7 feet	6 feet	6 feet
Bracing horizontal	1 x 4 in.	1 x 4 in.	1 x 6 in. or 1¼ x 4 in.	2 x 4 in.
Bracing diagonal	1 x 4 in.	1 x 4 in.	1 x 4 in.	2 x 4 in.
Tie-ins	1 x 4 in.	1 x 4 in.	1 x 4 in.	1 x 4 in.

Note: All members except planking are used on edge. All wood bearers shall be reinforced with $\frac{3}{16}$ x 2 inch steel strip, or the equivalent, secured to the lower edges for the entire length of the bearer.

MINIMUM SIZE OF MEMBERS – TUBE SCAFFOLDS

	Light duty	Medium duty	Heavy duty
Maximum intended load (lbs/ft^2)	25	50	75
Posts, runners and braces	Nominal 2 in. (1.90 inches) OD steel tube or pipe	Nominal 2 in. (1.90 inches) OD steel tube or pipe	Nominal 2 in. (1.90 inches) OD steel tube or pipe
Bearers	Nominal 2 in. (1.90 inches) OD steel tube or pipe and a maximum post spacing of 4 ft. x 10 ft.	Nominal 2 in. (1.90 inches) OD steel tube or pipe and a maximum post spacing of 4 ft. x 7 ft. or Nominal 2½ in. (2.375 in.) OD steel tube or pipe and a maximum post spacing of 6 ft. x 8 ft.*	Nominal 2½ in. (2.375 in.) OD steel tube or pipe and a maximum post spacing of 6 ft. x 6 ft.
Maximum runner spacing vertically	6 ft. 6 in.	6 ft. 6 in.	6 ft. 6 in.

*Bearers shall be installed in the direction of the shorter dimension.

Note: Longitudinal diagonal bracing shall be installed at an angle of 45 degrees (+/- 5 degrees).

SCHEDULE FOR LADDER-TYPE PLATFORMS

Length of Platform	12 ft.	14 and 16 ft.	18 and 20 ft.	22 and 24 ft.	28 and 30 ft.
Side stringers, minimum cross section (finished sizes):					
At ends	$1\frac{3}{4}$ x $2\frac{3}{4}$ in.	$1\frac{3}{4}$ x $2\frac{3}{4}$ in.	$1\frac{3}{4}$ x 3 in.	$1\frac{3}{4}$ x 3 in.	$1\frac{3}{4}$ x $3\frac{1}{2}$ in.
At middle	$1\frac{3}{4}$ x $3\frac{3}{4}$ in.	$1\frac{3}{4}$ x $3\frac{3}{4}$ in.	$1\frac{3}{4}$ x 4 in.	$1\frac{3}{4}$ x $4\frac{1}{4}$ in.	$1\frac{3}{4}$ x 5 in.
Reinforcing strip (minimum)	A $\frac{1}{8}$ x $\frac{7}{8}$-inch steel reinforcing strip shall be attached to the side or underside, full length.				
Rungs	Rungs shall be $1\frac{1}{8}$ inch minimum diameter with at least $\frac{7}{8}$ inch in diameter tenons, and the maximum spacing shall be 12 inches to center.				
Tie rods Number (minimum)	3	4	4	5	6
Diameter (minimum)	$\frac{1}{4}$ in.	$\frac{1}{4}$ in.	$\frac{1}{4}$ in.	$\frac{1}{4}$ in.	$\frac{1}{4}$ in.
Flooring, minimum finished size	$\frac{1}{2}$ x $2\frac{3}{4}$ in.	$\frac{1}{2}$ x $2\frac{3}{4}$ in.	$\frac{1}{2}$ x $2\frac{3}{4}$ in.	$\frac{1}{2}$ x $2\frac{3}{4}$ in.	$\frac{1}{2}$ x $2\frac{3}{4}$ in.

SCAFFOLD LOADS

Rated load capacity	Intended load
Light-duty	25 pounds per square foot applied uniformly over the entire span area.
Medium-duty	50 pounds per square foot applied uniformly over the entire span area.
Heavy-duty	75 pounds per square foot applied uniformly over the entire span area.
One-person	250 pounds placed at the center of the span (total 250 pounds).
Two-person	250 pounds placed 18 inches to the left and right of the center of the span (total 500 pounds).
Three-person	250 pounds placed at the center of the span and 250 pounds placed 18 inches to the left and right of the center of the span (total 750 pounds).

Note: Platform units used to make scaffold platforms intended for light-duty use shall be capable of supporting at least 25 pounds per square foot applied uniformly over the entire unit-span area, or a 250-pound point load placed on the unit at the center of the span, whichever load produces the greater shear force.

ALLOWABLE SCAFFOLD SPANS

2 x 10 inch (nominal) or 2 x 9 inch (rough) solid sawn wood planks

Maximum intended nominal load (lb/ft²)	Maximum permissible span using full thickness undressed lumber (ft)	Maximum permissible span using nominal thickness lumber (ft)
25	10	8
50	8	6
75	6	4

MAXIMUM NUMBER OF PLANKED LEVELS

Number of Working Levels	Maximum number of additional planked levels			Maximum height of scaffold (in feet)
	Light duty	Medium duty	Heavy duty	
1	16	11	6	125
2	11	1	0	125
3	6	0	0	125
4	1	0	0	125

SAFETY NETS
Outward from the outermost projection of the work surface

Vertical distance from working level to horizontal plane of net	Minimum required horizontal distance of outer edge of net from the edge of the working surface
Up to 5 feet	8 feet
More than 5 feet, up to 10 feet	10 feet
More than 10 feet	13 feet

CHAPTER 2
Excavation

SOIL TYPES

Type A soils are cohesive soils with an unconfined, compressive strength of 1.5 tons per square foot (tsf) (144kPa) or greater. Examples of Type A soils include clay, silty clay, sandy clay, clay loam and, in some cases, silty clay loam and sandy clay loam. Cemented soils such as caliche and hardpan are also considered Type A. However, no soil is Type A if:

- The soil is fissured.
- The soil is subject to vibration from heavy traffic, pile driving or similar effects.
- The soil has been previously disturbed.
- The soil is part of a sloped, layered system where the layers dip into the excavation on a slope of 4 horizontal to 1 vertical (4H:1V) or greater.
- The material is subject to other factors that would require it to be classified as a less stable material.

Type B soils are cohesive soils with an unconfined, compressive strength greater than 0.5 tsf (48 kPa) but less than 1.5 tsf (144 kPa). Examples of Type B soils include angular gravel (similar to crushed rock), silt, silt loam, sandy loam and, in some cases, silty clay loam and sandy clay loam; previously disturbed soils

SOIL TYPES *(cont.)*

except those which would otherwise be classified as Type C soils; soils that meet the unconfined, compressive strength or cementation requirements for Type A, but are fissured or subject to vibration; dry rock that is not stable, and material that is part of a sloped, layered system where the layers dip into the excavation on a slope less steep than 4 horizontal to 1 vertical (4H:1V), but only if the material would otherwise be classified as Type B.

Type C soils are cohesive soils with an unconfined, compressive strength of 0.5 tsf (48 kPa) or less. Examples of Type C soils include granular soils such as gravel, sand and loamy sand; submerged soil or soil from which water is freely seeping; submerged rock that is not stable, and material in a sloped, layered system where the layers dip into the excavation or a slope of 4 horizontal to 1 vertical (4H:1V) or steeper.

Unconfined compressive strength means the load per unit area at which a soil will fail in compression. It can be determined by laboratory testing or estimated in the field using a pocket penetrometer or by thumb penetration tests and other methods.

Wet soil means soil that contains significantly more moisture than moist soil, but in such a range of values that cohesive material will slump or begin to flow when vibrated. Granular material that would exhibit cohesive properties when moist will lose those cohesive properties when wet.

TERMS USED IN GRADING

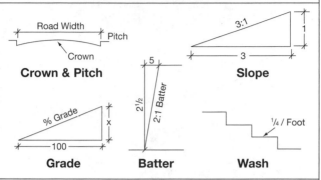

Crown & Pitch

Slope

Grade

Batter

Wash

MAXIMUM ALLOWABLE SLOPES

Soil or Rock Type	Maximum Allowable Slopes (H:V)[1] for Excavations Less Than 20 Feet Deep[3]
Stable Rock	Vertical (90 degrees)
Type A[2]	¾:1 (53 degrees)
Type B	1:1 (45 degrees)
Type C	1½:1 (34 degrees)

[1]Numbers shown in parentheses next to maximum allowable slopes are angles expressed in degrees from the horizontal. Angles have been rounded off.

[2]A short-term maximum allowable slope of ½H:1V (63 degrees) is allowed in excavations in Type A soil that are 12 feet (3.67 m) or less in depth. Short-term maximum allowable slopes for excavations greater than 12 feet (3.67 m) in depth shall be ¾H:1V (53 degrees).

[3]Sloping or benching for excavations greater than 20 feet deep should be designed by a registered professional engineer.

EXCAVATING TOWARD AN OPEN DITCH

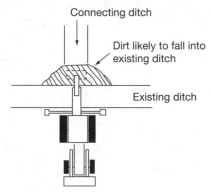

Connecting ditch

Dirt likely to fall into existing ditch

Existing ditch

Hazard From Falling Dirt

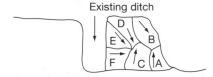

Existing ditch

Excavate in This Sequence to Avoid Falling Dirt

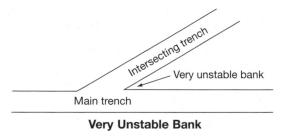

Intersecting trench

Very unstable bank

Main trench

Very Unstable Bank

SEQUENCE OF EXCAVATION IN HIGH GROUNDWATER

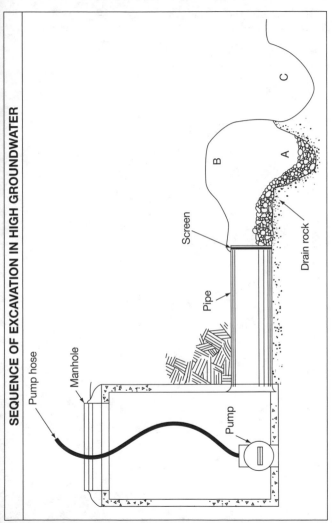

Pump hose

Manhole

Screen

Pipe

Drain rock

Pump

A

B

C

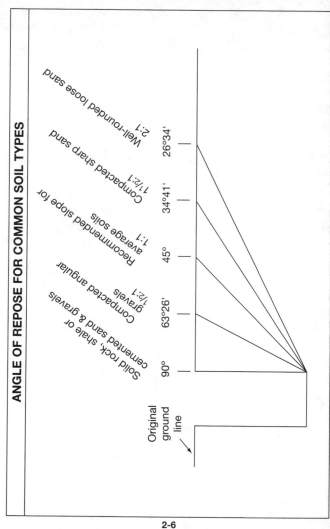

ANGLE OF REPOSE FOR COMMON SOIL TYPES

Well-rounded loose sand
2:1
26°34'

Compacted sharp sand
1½:1
34°41'

Recommended slope for average soils
1:1
45°

Compacted angular gravels
½:1
63°26'

Solid rock, shale or cemented sand & gravels
90°

Original ground line

2-6

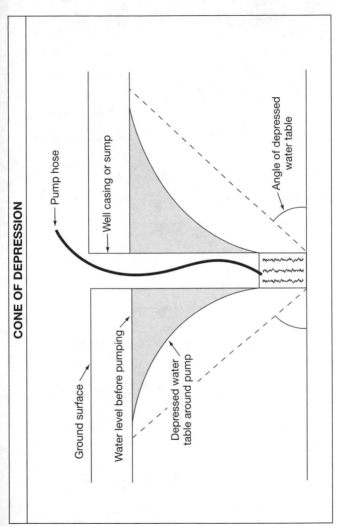

CONE OF DEPRESSION

Pump hose

Well casing or sump

Angle of depressed water table

Ground surface

Water level before pumping

Depressed water table around pump

Angle of depressed water table

2-7

EXCAVATIONS MADE IN TYPE A SOIL

All simple slope excavations 20' or less in depth shall have a maximum allowable slope of ¾:1.

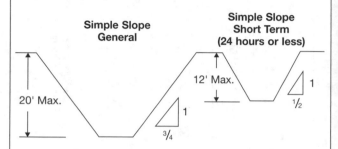

Simple Slope General

Simple Slope Short Term (24 hours or less)

Exception: Simple slope excavations which are open 24 hours or less (short term) and which are 12' or less in depth shall have a maximum allowable slope of ½:1.

All benched excavations 20' or less in depth shall have a maximum allowable slope of ¾ to 1 and maximum bench dimensions as follows:

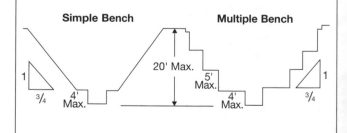

Simple Bench

Multiple Bench

EXCAVATIONS MADE IN TYPE B SOIL

All simple slope excavations 20' or less in depth shall have a maximum allowable slope of 1:1.

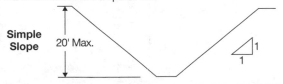

Simple Slope

20' Max.

All benched excavations 20' or less in depth shall have a maximum allowable slope of 1:1 and maximum bench dimensions as follows:

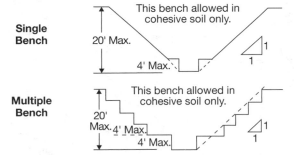

Single Bench

This bench allowed in cohesive soil only.

20' Max.

4' Max.

Multiple Bench

This bench allowed in cohesive soil only.

20' Max.

4' Max.

4' Max.

All excavations 20' or less in depth which have vertically sided lower portions shall be shielded or supported to a height at least 18" above the top of the vertical side. All such excavations shall have a maximum allowable slope of 1:1.

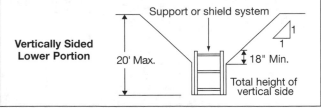

Vertically Sided Lower Portion

Support or shield system

20' Max.

18" Min.

Total height of vertical side

EXCAVATIONS MADE IN TYPE C SOIL

All simple slope excavations 20' or less in depth shall have a maximum allowable slope of 1½:1.

Simple Slope

All excavations 20' or less in depth which have vertically sided lower portions shall be shielded or supported to a height at least 18" above the top of the vertical side. All such excavations shall have a maximum allowable slope of 1½:1.

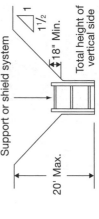

Vertically Sided Lower Portion

All excavations 8' or less in depth which have unsupported vertically sided lower portions shall have a maximum vertical side of 3½'.

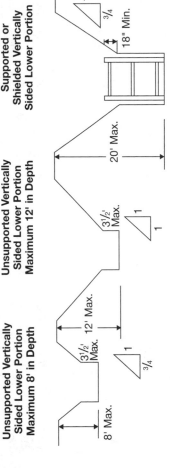

Unsupported Vertically Sided Lower Portion Maximum 8' in Depth

8' Max.

3½' Max.

¾ / 1

Unsupported Vertically Sided Lower Portion Maximum 12' in Depth

12' Max.

3½' Max.

1 / 1

20' Max.

Supported or Shielded Vertically Sided Lower Portion

18" Min.

¾ / 1

All excavations more than 8' but not more than 12' in depth with unsupported vertically sided lower portions shall have a maximum allowable slope of 1:1 and a maximum vertical side of 3½'.

All excavations 20' or less in depth which have vertically sided lower portions that are supported or shielded shall have a maximum allowable slope of ¾:1. The support or shield system must extend at least 18" above the top of the vertical side.

EXCAVATIONS MADE IN LAYERED SOILS

All excavations 20' or less in depth made in layered soils have a maximum allowable slope for each layer.

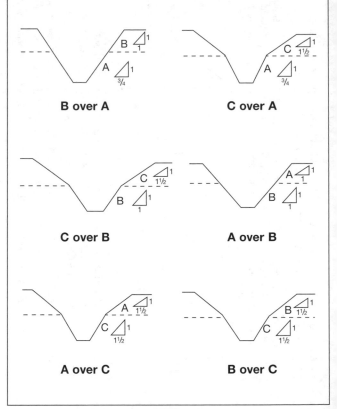

B over A

C over A

C over B

A over B

A over C

B over C

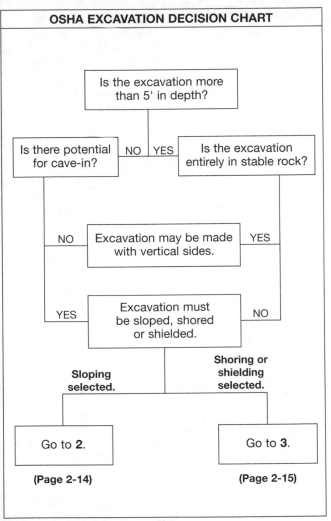

OSHA EXCAVATION DECISION CHART

Is the excavation more than 5' in depth?

NO | YES

Is there potential for cave-in?

Is the excavation entirely in stable rock?

NO | Excavation may be made with vertical sides. | YES

YES | Excavation must be sloped, shored or shielded. | NO

Sloping selected.

Shoring or shielding selected.

Go to **2**.

Go to **3**.

(Page 2-14)

(Page 2-15)

OSHA EXCAVATION DECISION CHART *(cont.)*

2
Sloping selected as the method of protection.

Will soil classification be made in accordance with §1926.652 (b)?

YES

NO

Excavation must comply with one of the following three options:

Option 1:
§1926.652 (b)(2) which requires Appendices A and B to be followed.

Option 2:
§1926.652 (b)(3) which requires other tabulated data to be followed.

Option 3:
§1926.652 (b)(4) which requires the excavation to be designed by a registered professional engineer.

Excavations must comply with §1926.652 (b)(1) which requires a slope of 1½H:1V (34°).

3
Shoring or shielding selected as the method of protection.

Soil classification is required when shoring or shielding is used. The excavation must comply with one of the following four options:

Option 1:
§1926.652 (c)(1) which requires Appendices A and C to be followed (e.g., timber shoring).

Option 2:
§1926.652 (c)(2) which requires manufacturers data to be followed (e.g., hydraulic shoring, trench jacks, air shores, shields).

Option 3:
§1926.652 (c)(3) which requires tabulated data to be followed (e.g., any system as per the tabulated data).

Option 4:
§1926.652 (c)(4) which requires the excavation to be designed by a registered professional engineer (e.g., any designed system).

TIMBER TRENCH SHORING
MINIMUM TIMBER REQUIREMENTS — SOIL TYPE A

$P(a) = 25 \times H + 72$ psf (2 ft. surcharge)

Depth of Trench (feet)	Size (actual) and Spacing of Members						
	Cross Braces						
	Horiz. Spacing (feet)	Width of Trench (feet)					Vertical Spacing (feet)
		Up to 4	Up to 6	Up to 9	Up to 12	Up to 15	
5 to 10	Up to 6	4 x 4	4 x 4	4 x 6	6 x 6	6 x 6	4
	Up to 8	4 x 4	4 x 4	4 x 6	6 x 6	6 x 6	4
	Up to 10	4 x 6	4 x 6	4 x 6	6 x 6	6 x 6	4
	Up to 12	4 x 6	4 x 6	6 x 6	6 x 6	6 x 6	4
10 to 15	Up to 6	4 x 4	4 x 4	4 x 6	6 x 6	6 x 6	4
	Up to 8	4 x 6	4 x 6	6 x 6	6 x 6	6 x 6	4
	Up to 10	6 x 6	6 x 6	6 x 6	6 x 8	6 x 8	4
	Up to 12	6 x 6	6 x 6	6 x 6	6 x 8	6 x 8	4
15 to 20	Up to 6	6 x 6	6 x 6	6 x 6	6 x 8	6 x 8	4
	Up to 8	6 x 6	6 x 6	6 x 6	6 x 8	6 x 8	4
	Up to 10	8 x 8	8 x 8	8 x 8	8 x 8	8 x 10	4
	Up to 12	8 x 8	8 x 8	8 x 8	8 x 8	8 x 10	4

TIMBER TRENCH SHORING *(cont.)*
MINIMUM TIMBER REQUIREMENTS — SOIL TYPE A

P(a) = 25 x H + 72 psf (2 ft. surcharge)

Depth of Trench (feet)	Size (actual) and Spacing of Members						
	Wales		Uprights				
		Vertical	Maximum Allowable Horizontal Spacing (feet)				
	Size (in.)	Spacing (feet)	Close	4	5	6	8
5 to 10	Not req'd.	–	–	–	–	2 x 6	–
	Not req'd.	–	–	–	–	–	2 x 8
	8 x 8	4	–	–	2 x 6	–	–
	8 x 8	4	–	–	–	2 x 6	–
10 to 15	Not req'd.	–	–	–	–	3 x 8	–
	8 x 8	4	–	2 x 6	–	–	–
	8 x 10	4	–	–	2 x 6	–	–
	10 x 10	4	–	–	–	3 x 8	–
15 to 20	6 x 8	4	3 x 6	–	–	–	–
	8 x 8	4	3 x 6	–	–	–	–
	8 x 10	4	3 x 6	–	–	–	–
	10 x 10	4	3 x 6	–	–	–	–

Mixed oak or equivalent with a bending strength not less than 850 psi. Manufactured members of equivalent strength may be substituted for wood.

TIMBER TRENCH SHORING
MINIMUM TIMBER REQUIREMENTS — SOIL TYPE B

P(a) = 45 x H + 72 psf (2 ft. surcharge)

Depth of Trench (feet)	Horiz. Spacing (feet)	Cross Braces					Vertical Spacing (feet)
		Width of Trench (feet)					
		Up to 4	Up to 6	Up to 9	Up to 12	Up to 15	
5 to 10	Up to 6	4 x 6	4 x 6	6 x 6	6 x 6	6 x 6	5
	Up to 8	6 x 6	6 x 6	6 x 6	6 x 8	6 x 8	5
	Up to 10	6 x 6	6 x 6	6 x 6	6 x 8	6 x 8	5
10 to 15	Up to 6	6 x 6	6 x 6	6 x 6	6 x 8	6 x 8	5
	Up to 8	6 x 8	6 x 8	6 x 8	8 x 8	8 x 8	5
	Up to 10	8 x 8	8 x 8	8 x 8	8 x 8	8 x 10	5
15 to 20	Up to 6	6 x 8	6 x 8	6 x 8	8 x 8	8 x 8	5
	Up to 8	8 x 8	8 x 8	8 X 8	8 x 8	8 x 10	5
	Up to 10	8 x 10	8 x 10	8 X 10	8 x 10	10 x 10	5

TIMBER TRENCH SHORING *(cont.)*
MINIMUM TIMBER REQUIREMENTS — SOIL TYPE B

P(a) = 45 x H + 72 psf (2 ft. surcharge)

Depth of Trench (feet)	Size (actual) and Spacing of Members				
	Wales		Uprights		
	Size (in.)	Vertical Spacing (feet)	Maximum Allowable Horizontal Spacing (feet)		
			Close	2	3
5 to 10	6 x 8	5	–	–	2 x 6
	8 x 10	5	–	–	2 x 6
	10 x 10	5	–	–	2 x 6
10 to 15	8 x 8	5	–	2 x 6	–
	10 x 10	5	–	2 x 6	–
	10 x 12	5	–	2 x 6	–
15 to 20	8 x 10	5	3 x 6	–	–
	10 x 12	5	3 x 6	–	–
	12 x 12	5	3 x 6	–	–

Mixed oak or equivalent with a bending strength not less than 850 psi. Manufactured members of equivalent strength may be substituted for wood.

TIMBER TRENCH SHORING
MINIMUM TIMBER REQUIREMENTS — SOIL TYPE C

$P(a) = 80 \times H + 72$ psf (2 ft. surcharge)

Depth of Trench (feet)	Horiz. Spacing (feet)	Cross Braces					Vertical Spacing (feet)
		Width of Trench (feet)					
		Up to 4	Up to 6	Up to 9	Up to 12	Up to 15	
5 to 10	Up to 6	6 x 8	6 x 8	6 x 8	8 x 8	8 x 8	5
	Up to 8	8 x 8	8 x 8	8 x 8	8 x 8	8 x 10	5
	Up to 10	8 x 10	8 x 10	8 x 10	8 x 10	10 x 10	5
10 to 15	Up to 6	8 x 8	8 x 8	8 x 8	8 x 8	8 x 10	5
	Up to 8	8 x 10	8 x 10	8 x 10	8 x 10	10 x 10	5
15 to 20	Up to 6	8 x 10	8 x 10	8 x 10	8 x 10	10 x 10	5

TIMBER TRENCH SHORING (cont.)
MINIMUM TIMBER REQUIREMENTS — SOIL TYPE C

$P(a) = 80 \times H + 72$ psf (2 ft. surcharge)

| Depth of Trench (feet) | Size (actual) and Spacing of Members | | Uprights |
| | Wales | | Maximum Allowable Horizontal Spacing (feet) |
	Size (in.)	Vertical Spacing (feet)	Close
5 to 10	8 x 10	5	2 x 6
	10 x 12	5	2 x 6
	12 x 12	5	2 x 6
10 to 15	10 x 12	5	2 x 6
	12 x 12	5	2 x 6
15 to 20	12 x 12	5	3 x 6

Mixed oak or equivalent with a bending strength not less than 850 psi. Manufactured members of equivalent strength may be substituted for wood.

TIMBER TRENCH SHORING
MINIMUM TIMBER REQUIREMENTS — SOIL TYPE A

P(a) = 25 x H + 72 psf (2 ft. surcharge)

Depth of Trench (feet)	Size (S4S) and Spacing of Members						
	Cross Braces						
	Horiz. Spacing (feet)	Width of Trench (feet)					Vertical Spacing (feet)
		Up to 4	Up to 6	Up to 9	Up to 12	Up to 15	
5 to 10	Up to 6	4 x 4	4 x 4	4 x 4	4 x 4	4 x 6	4
	Up to 8	4 x 4	4 x 4	4 x 4	4 x 6	4 x 6	4
	Up to 10	4 x 6	4 x 6	4 x 6	6 x 6	6 x 6	4
	Up to 12	4 x 6	4 x 6	4 x 6	6 x 6	6 x 6	4
10 to 15	Up to 6	4 x 4	4 x 4	4 x 4	6 x 6	6 x 6	4
	Up to 8	4 x 6	4 x 6	4 x 6	6 x 6	6 x 6	4
	Up to 10	6 x 6	6 x 6	6 x 6	6 x 6	6 x 6	4
	Up to 12	6 x 6	6 x 6	6 x 6	6 x 6	6 x 6	4
15 to 20	Up to 6	6 x 6	6 x 6	6 x 6	6 x 6	6 x 6	4
	Up to 8	6 x 6	6 x 6	6 x 6	6 x 6	6 x 6	4
	Up to 10	6 x 6	6 x 6	6 x 6	6 x 6	6 x 8	4
	Up to 12	6 x 6	6 x 6	6 x 6	6 x 8	6 x 8	4

TIMBER TRENCH SHORING *(cont.)*
MINIMUM TIMBER REQUIREMENTS — SOIL TYPE A

P(a) = 25 x H + 72 psf (2 ft. surcharge)

Depth of Trench (feet)	Size (S4S) and Spacing of Members						
	Wales		Uprights				
			Maximum Allowable Horizontal Spacing (feet)				
	Size (in.)	Vertical Spacing (feet)	Close	4	5	6	8
5 to 10	Not req'd.	Not req'd.	–	–	–	4 x 6	–
	Not req'd.	Not req'd.	–	–	–	–	4 x 8
	8 x 8	4	–	–	4 x 6	–	–
	8 x 8	4	–	–	–	4 x 6	–
10 to 15	Not req'd.	Not req'd.	–	–	–	4 x 10	–
	6 x 8	4	–	4 x 6	–	–	–
	8 x 8	4	–	–	4 x 8	–	–
	8 x 10	4	–	4 x 6	–	4 x 10	–
15 to 20	6 x 8	4	3 x 6	–	–	–	–
	8 x 8	4	3 x 6	4 x 12	–	–	–
	8 x 10	4	3 x 6	–	–	–	–
	8 x 12	4	3 x 6	4 x 12	–	–	–

Douglas fir or equivalent with a bending strength not less than 1500 psi. Manufactured members of equivalent strength may be substituted for wood.

TIMBER TRENCH SHORING
MINIMUM TIMBER REQUIREMENTS — SOIL TYPE B

$$P(a) = 45 \times H + 72 \text{ psf (2 ft. surcharge)}$$

Depth of Trench (feet)	Horiz. Spacing (feet)	Up to 4	Up to 6	Up to 9	Up to 12	Up to 15	Vertical Spacing (feet)
		Size (S4S) and Spacing of Members					
		Cross Braces					
		Width of Trench (feet)					
5 to 10	Up to 6	4 x 6	4 x 6	4 x 6	6 x 6	6 x 6	5
	Up to 8	4 x 6	4 x 6	6 x 6	6 x 6	6 x 6	5
	Up to 10	4 x 6	4 x 6	6 x 6	6 x 6	6 x 8	5
10 to 15	Up to 6	6 x 6	6 x 6	6 x 6	6 x 8	6 x 8	5
	Up to 8	6 x 8	6 x 8	6 x 8	8 x 8	8 x 8	5
	Up to 10	6 x 8	6 x 8	8 x 8	8 x 8	8 x 8	5
15 to 20	Up to 6	6 x 8	6 x 8	6 x 8	6 x 8	8 x 8	5
	Up to 8	6 x 8	6 x 8	6 x 8	8 x 8	8 x 8	5
	Up to 10	8 x 8	8 x 8	8 x 8	8 x 8	8 x 8	5

TIMBER TRENCH SHORING *(cont.)*
MINIMUM TIMBER REQUIREMENTS — SOIL TYPE B

P(a) = 45 x H + 72 psf (2 ft. surcharge)

Depth of Trench (feet)	Size (S4S) and Spacing of Members				
	Wales		Uprights		
	Size (in.)	Vertical Spacing (feet)	Maximum Allowable Horizontal Spacing (feet)		
			Close	2	3
5 to 10	6 x 8	5	–	–	3 x 12 4 x 8
	8 x 8	5	–	3 x 8	
	8 x 10	5	–	–	4 x 8
10 to 15	8 x 8	5	3 x 6	4 x 10	–
	10 x 10	5	3 x 6	4 x 10	–
	10 x 12	5	3 x 6	4 x 10	–
15 to 20	8 x 10	5	4 x 6	–	–
	10 x 12	5	4 x 6	–	–
	12 x 12	5	4 x 6	–	–

Douglas fir or equivalent with a bending strength not less than 1500 psi. Manufactured members of equivalent strength may be substituted for wood.

TIMBER TRENCH SHORING
MINIMUM TIMBER REQUIREMENTS — SOIL TYPE C

$P(a) = 80 \times H + 72$ psf (2 ft. surcharge)

Depth of Trench (feet)	Size (S4S) and Spacing of Members						
	Cross Braces						
	Horiz. Spacing (feet)	Width of Trench (feet)					Vertical Spacing (feet)
		Up to 4	Up to 6	Up to 9	Up to 12	Up to 15	
5 to 10	Up to 6	6 x 6	6 x 6	6 x 6	6 x 6	8 x 8	5
	Up to 8	6 x 6	6 x 6	6 x 6	8 x 8	8 x 8	5
	Up to 10	6 x 6	6 x 6	8 x 8	8 x 8	8 x 8	5
10 to 15	Up to 6	6 x 8	6 x 8	6 x 8	8 x 8	8 x 8	5
	Up to 8	8 x 8	8 x 8	8 x 8	8 x 8	8 x 8	5
15 to 20	Up to 6	8 x 8	8 x 8	8 x 8	8 x 10	8 x 10	5

TIMBER TRENCH SHORING (cont.)
MINIMUM TIMBER REQUIREMENTS — SOIL TYPE C

P(a) = 80 x H + 72 psf (2 ft. surcharge)

Depth of Trench (feet)	Size (S4S) and Spacing of Members		Uprights
	Wales		
	Size (in.)	Vertical Spacing (feet)	Maximum Allowable Horizontal Spacing (feet)
			Close
5 to 10	8 x 8	5	3 x 6
	10 x 10	5	3 x 6
	10 x 12	5	3 x 6
10 to 15	10 x 10	5	4 x 6
	12 x 12	5	4 x 6
15 to 20	10 x 12	5	4 x 6

LOAD-CARRYING CAPACITIES OF SOILS

Type of soil	Tons per sq. ft.
Soft clay	1
Firm clay or fine sand	2
Compact fine or loose coarse sand	3
Loose gravel or compact coarse sand	4
Compact sand-gravel mixture	6

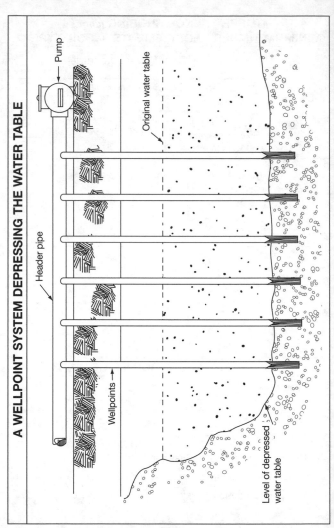

A WELLPOINT SYSTEM DEPRESSING THE WATER TABLE

Pump

Header pipe

Original water table

Wellpoints

Level of depressed water table

2-28

CHAPTER 3
Concrete and Masonry

ORDERING READY-MIX CONCRETE	
Kind of Work	**Order the Following***

Bond beams
Chimney caps
Lintels
Reinforced concrete beams,
    girders and other sections
Reinforced concrete floors,
    roof slabs and top courses
Septic tanks
	A mix containing at least $6\frac{1}{2}$ sacks of portland cement per cubic yard and a maximum of 6 gallons of water per sack of cement (using $\frac{3}{4}$" maximum size aggregate).

Basement floors Curbs Driveways Entrance platforms and steps Garage floors Patio slabs Sidewalks Slabs on ground Stairs Swimming pools	A mix containing at least 6 sacks of portland cement per cubic yard and a maximum of 6 gallons of water per sack of cement.
Footings Foundation walls Retaining walls	A mix containing at least 5 sacks of portland cement per cubic yard and a maximum of 7 gallons of water per sack of cement.

*Order air-entrained concrete for all concrete exposed to freezing and thawing.

RECOMMENDED WATER/CEMENT RATIO FOR DIFFERENT APPLICATIONS AND CONDITIONS

Exposure	Class of Structure		
	Reinforced piles, thin walls, light structural members, exterior columns and beams in buildings	Reinforced reservoirs, water tanks, pressure pipes, sewers, canal linings, dams of thin sections	Heavy walls, piers, foundations, dams of heavy sections
	Water/Cement Ratio, U.S. Gallons per Sack[1]		
Extreme: In severe climates like northern U.S., exposure to rain and snow and drying, freezing and thawing, as at the water line in hydraulic structures. Exposure to sea and strong sulphate waters in both severe and moderate climates.	5½	5½	6

Exposure condition			
Severe:			
In severe climates like northern U.S., exposure to rain and snow and freezing and thawing, but not continuously in contact with water.	6	6	6¾
In moderate climates like southern U.S., exposure to alternate wetting and drying, as at the water line in hydraulic structures.			
Moderate:			
In climates like southern U.S., exposure to ordinary weather, but not continuously in contact with water.	6¾	6	7½
Concrete completely submerged, but protected from freezing.	7½		
Protected:			
Ordinary inclosed structural members; concrete below the ground and not subject to action of corrosive groundwaters or freezing and thawing.		6	8¼

3-3

[1]Surface water or moisture carried by the aggregate must be included as part of the mixing water.

APPROXIMATE RELATIVE STRENGTH OF CONCRETE AS AFFECTED BY TYPE OF CEMENT

Type of Portland Cement		Compressive Strength, Percent of Strength of Type 1 or Normal Portland Cement Concrete			
ASTM	CSA	1 day	7 days	28 days	3 months
I	Normal	100	100	100	100
II		75	85	90	100
III	High-Early-Strength	190	120	110	100
IV		55	55	75	100
V	Sulfate-Resistant	65	75	85	100

VOLUME OF CONCRETE MIXTURES

	Proportions of Mixture		
Cement (bag)	Sand (cu. ft.)	Gravel or Stone (cu. ft.)	Volume of Concrete (cu. ft.)
1	1½	3	3½
1	2	3	3⁹/₁₀
1	2	4	4½
1	2½	5	5²/₅
1	3	5	5⁴/₅

SPECIFIC GRAVITIES AND UNIT WEIGHTS

Material	Specific Gravity	Unit Weight in Pounds Per Cubic Foot
Portland cement	3.15	94
Lime	2.25	40
Water	1.00	62.4

Note: Values for sand vary considerably.

WATER USAGE

If Mix Calls For:	Use These Amounts of Mixing Water, in Gallons, When Sand is:		
	Damp	Wet	Very Wet
6 gal. per sack of cement	5½	5	4¼
7 gal. per sack of cement	6¼	5½	4¾

RECOMMENDED PROPORTIONS FOR FLAT CONCRETE WORK

Portland Cement (94 lb. sacks) (1 cu. ft.)*	Water for Sand That is:			Sand	Aggregate
	Very Wet	Average Wet	Damp		
For 1 sack portland cement 1 sack	4¼ gals.	5 gals.	5½ gals.	2¼ cu. ft.	3 cu. ft.
For 1 cu. yd. concrete (27 cu. ft.) 6¼ sacks	25½ gals.	30 gals.	38 gals.	1400 lbs.	2100 lbs.

*One sack of cement makes about 4½ cu. ft. of concrete.

MIXING SMALL AMOUNTS OF CONCRETE

Concrete Required (cu. ft.)	Cement* (lbs.)	Maximum Amount of Water to Use (gallons)		Sand (lbs.)	Coarse Aggregate (lbs.)
		U.S.	Imperial		
1	24	1¼	1	52	78
3	71	3¾	3⅛	156	233
5	118	6¼	5¼	260	389
6¾ (¼ cu. yd.)	165	8	6¾	350	525
13½ (½ cu. yd.)	294	16	13½	700	1,050
27 (1 cu. yd.)	588	32	27	1,400	2,100

*U.S. bag of cement weighs 94 lbs. Canadian bag of cement weighs 80 lbs.
A 1:2¼:3 mix = 1 part cement to 2¼ parts sand to 3 parts (1-in. max.) aggregate.

3-7

RECOMMENDED PROPORTIONS OF WATER TO CEMENT

Kinds of Work	Add U.S. Gallons of Water to Each Sack Batch if Sand is:			Suggested Mixture for Trial Batch			Materials per cu. yd. of Concrete*		
	Very Wet	Wet	Damp	Cement Sacks	Aggregates Fine Cu. Ft.	Aggregates Coarse Cu. Ft.	Cement Sacks	Aggregates Fine Cu. Ft.	Aggregates Coarse Cu. Ft.
5-Gallon Paste for Concrete Subjected to Severe Wear, Weather or Weak Acid and Alkali Solutions									
Colored or plain topping for heavy wearing surfaces as in industrial plants and all other two-course work such as pavements, walks, tennis courts, residence floors, etc.	4¼	Average Sand 4¼	4¾	1	1	1¾	10	10	17
					Maximum size aggregate ⅜"				
One-course industrial, creamery and dairy plant floors and all other concrete in contact with weak acid or alkali solutions.	3¾	Average Sand 4	4½	1	1¾	2	8	14	16
					Maximum size aggregate ¾"				

3-8

6-Gallon Paste for Concrete to be Watertight or Subjected to Moderate Wear and Weather

Watertight floors such as industrial plant, basement, dairy barn, etc. Watertight foundations.

Concrete subjected to moderate wear or frost action such as driveways, walks, tennis courts, etc.

All watertight concrete for swimming and wading pools, septic tanks, storage tanks, etc.

All base course work such as floors, walks, etc.

All reinforced concrete structural beams, columns, slabs, residence floors, etc.

	Average Sand							
4¼	5	5½	1	2¼	3	6¼	14	19

Maximum size aggregate 1½"

7-Gallon Paste for Concrete not Subjected to Wear, Weather or Water

Foundation walls, footings, mass concrete, etc., not subjected to weather, water pressure or other exposure.

	Average Sand							
4¾	5½	6¼	1	2¾	4	5	14	20

Maximum size aggregate 1½"

EFFECT ON FINAL CURE STRENGTH OF CONCRETE WHEN THE CEMENT/WATER RATIO IS CHANGED

*Type 1 portland cement, constant proportions of cement, sand and coarse aggregate. Moist cured at 70°F for 28 days.

3-10

MATERIALS REQUIRED FOR SMALL QUANTITIES OF CONCRETE

Cubic Feet of Concrete	1:1½:3 Mixture			1:2:3 Mixture			1:2:4 Mixture			1:2½:5 Mixture			1:3:6 Mixture		
	Bags Cement	Cu. Ft. Sand	Cu. Ft. Stone	Bags Cement	Cu. Ft. Sand	Cu. Ft. Stone	Bags Cement	Cu. Ft. Sand	Cu. Ft. Stone	Bags Cement	Cu. Ft. Sand	Cu. Ft. Stone	Bags Cement	Cu. Ft. Sand	Cu. Ft. Stone
100	28	42	84	25 4/5	51 3/5	77 2/5	22	44	88	18	45	90	16	48	96
90	25 1/5	37 4/5	75 3/5	23 1/5	46 2/5	69 3/5	19 4/5	39 3/5	79 1/5	16 1/5	40 1/2	81	14 2/5	43 1/5	86 2/5
80	22 2/5	33 3/5	67 1/5	20 2/3	41 1/3	62	17 3/5	37 1/5	70 2/5	14 2/5	36	72	12 4/5	38 2/5	76 4/5
70	19 3/5	29 2/5	58 4/5	18	36	54	15 2/5	30 4/5	61 3/5	12 3/5	31 1/2	63	11 1/5	33 3/5	67 1/5
60	16 4/5	25 1/5	50 2/5	15 1/2	31	46 1/2	13 1/5	26 2/5	52 4/5	10 4/5	27	54	9 3/5	28 4/5	57 3/5
50	14	21	42	13	26	39	11	22	44	9	22 1/2	45	8	24	48
40	11 1/5	16 4/5	33 3/5	10 1/3	20 2/3	31	8 4/5	17 3/5	35 1/5	7 1/5	18	36	6 2/5	19 1/5	38 2/5
30	8 2/5	12 3/5	25 1/5	7 3/4	15 1/2	23 1/4	6 3/5	13 1/5	26 2/5	5 2/5	13 1/2	27	4 4/5	14 2/5	28 4/5
20	5 3/5	8 2/5	16 4/5	5 1/5	10 2/3	15 3/5	4 2/5	8 4/5	17 3/5	3 3/5	9	18	3 1/5	9 3/5	19 1/5
10	2 4/5	4 1/5	8 2/5	2 3/5	5 1/5	7 4/5	2 1/5	4 2/5	8 4/5	1 4/5	4 1/2	9	1 3/5	4 4/5	9 3/5
9	2 1/2	3 4/5	7 3/5	2 1/3	4 2/3	7	2	4	8	1 3/5	4	8	1 1/2	4 1/3	8 2/3
8	2 1/4	3 3/8	6 3/4	2	4 1/5	6 1/4	1 3/4	3 1/2	7	1 2/5	3 3/5	7 1/5	1 1/4	3 7/8	7 3/4
7	2	3	6	1 4/5	3 1/2	5 2/5	1 1/2	3	6	1 1/4	3 1/8	6 1/4	1 1/8	3 3/8	6 3/4
6	1 2/3	2 1/2	5	1 3/5	3 1/5	4 3/5	1 1/3	2 2/3	5 1/3	1 1/10	2 3/4	5 1/2	1	3	6
5	1 2/5	2 1/10	4 1/5	1 1/3	2 2/3	4	1 1/10	2 1/5	4 2/5	9/10	2 1/4	4 1/2	4/5	2 2/5	4 4/5
4	1 1/8	1 3/4	3 3/8	1	2	3 1/5	7/8	1 3/4	3 1/2	7/10	1 4/5	3 3/5	5/8	2	4
3	4/5	1 1/4	2 1/2	3/4	1 1/2	2 1/3	2/3	1 1/3	2 2/3	1/2	1 1/3	2 2/3	1/2	1 1/2	3
2	9/16	7/8	1 11/16	1/2	1	1 1/2	7/16	7/8	1 3/4	1/3	9/10	1 4/5	5/16	1	2
1	9/32	7/16	7/8	1/4	1/2	3/4	7/32	7/16	7/8	1/5	1/2	1	5/32	1/2	1

INGREDIENTS REQUIRED FOR 100 SQ. FT. OF CONCRETE

Thickness of Concrete (inches)	Amount of Concrete (cu. yd.)	Proportions*								
		1:2:2¼ Mix			1:2½:3½ Mix			1:3:4 Mix		
		Cement (sacks)	Aggregate		Cement (sacks)	Aggregate		Cement (sacks)	Aggregate	
			Fine (cu. ft.)	Coarse (cu. ft.)		Fine (cu. ft.)	Coarse (cu. ft.)		Fine (cu. ft.)	Coarse (cu. ft.)
3	0.92	7.1	14.3	16.1	5.5	13.8	19.3	4.6	13.8	18.4
4	1.24	9.6	19.2	21.7	7.4	18.6	26.0	6.2	18.6	24.8
5	1.56	12.1	24.2	27.3	9.4	23.4	32.8	7.8	23.4	31.2
6	1.85	14.3	28.7	32.4	11.1	27.8	38.9	9.3	27.8	37.0
8	2.46	19.1	38.1	43.0	14.8	36.9	51.7	12.3	36.9	49.3
10	3.08	23.9	47.7	53.9	18.5	46.2	64.7	15.4	46.2	61.6
12	3.70	28.7	57.3	64.7	22.2	55.5	77.7	18.5	55.5	74.0

*Quantities may vary 10% either way, depending on character of aggregate used. No allowance made in table for waste.

APPROXIMATE QUANTITIES OF MATERIALS REQUIRED FOR MAKING ONE CUBIC YARD OF CONCRETE

Proportions of the Concrete or Mortar			Quantities of Materials		
Cement	Sand	Gravel or Stone	Cement (sacks)	Sand (damp and loose) (cubic yards)	Gravel (loose) (cubic yards)
1	1.5	—	15.5	.86	—
1	2.0	—	12.8	.95	—
1	2.5	—	11.0	1.02	—
1	3.0	—	9.6	1.07	—
1	1.5	3	7.6	.42	.85
1	2.0	2	8.2	.60	.60
1	2.0	3	7.0	.52	.78
1	2.0	4	6.0	.44	.89
1	2.5	3.5	5.9	.55	.77
1	2.5	4	5.6	.52	.83
1	2.5	5	5.0	.46	.92
1	3.0	5	4.6	.51	.85
1	3.0	6	4.2	.47	.94

Note: These quantities are approximate and may vary by 10% depending on the aggregate used.

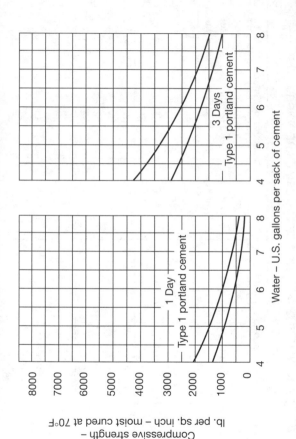

COMPRESSIVE STRENGTH IN PROPORTION TO WATER VOLUME

1 Day
Type 1 portland cement

3 Days
Type 1 portland cement

Compressive strength – lb. per sq. inch – moist cured at 70°F

Water – U.S. gallons per sack of cement

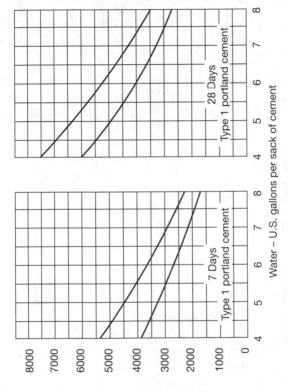

THE EFFECT OF CEMENT/WATER RATIOS

Compressive strength – lb. per sq. inch – moist cured at 70°F

Water – U.S. gallons per sack of cement

1 Day
Type III portland cement

3 Days
Type III portland cement

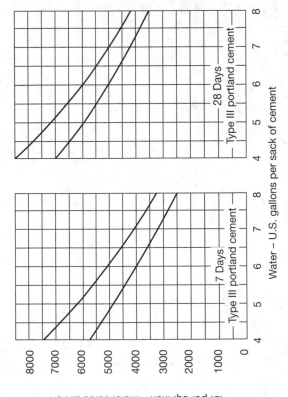

3-17

REQUIREMENTS FOR VARIOUS TYPES OF CONCRETE MASONRY UNITS

Specification, Serial Designation and Latest Revised Date	Minimum Faceshell Thickness (inches)	Compressive Strength, Minimum, psi, Average Gross Area		Water Absorption, Maximum, lb. per cu. ft. of Concrete, Average of 5 Units	Moisture Content, Maximum, Percent of Total Absorption, Average of 5 Units
		Average of 5 Units	Individual Unit		
Hollow load-bearing concrete masonry units	1¼ or over: Grade A[a, c] Grade B[b, c] Under 1¼ and over ¾	1000 700 1000	800 600 800	15 — 15	40 40 40
Hollow non-load-bearing concrete masonry units	Not less than ½	350	300	—	40

3-18

Solid load-bearing concrete masonry units					
Grade A	—	1800	1600	15	40
Grade B	—	1200	1000	15	40
				Average of 3 units	Average of 3 units
Concrete units; masonry, hollow					
Load-bearing units	1¼ or more ¾ to 1¼	700 1000	600 800	16 16	40 40
Non-load-bearing units	Not less than ¾	350	—	—	40

3-19

REQUIREMENTS FOR VARIOUS TYPES OF CONCRETE MASONRY UNITS (cont.)

	Compressive Strength, Minimum, psi, Average Gross Area (brick flatwise)		Modulus of Rupture, Minimum, psi, (brick flatwise)		Water Absorption, Maximum		Moisture Content, Maximum, Percent of Total Absorption
	Average of 5 Brick	Individual	Average of 5 Brick	Individual			
Concrete building brick					15 lb. per cu. ft.		
Grade A[e]	2500	2000	—	—			40
Grade B[f]	1500	1250	—	—			40
					Average of 5 Brick	Individual	
Brick; concrete							
H-Hard	—	—	600+	400	7 oz.	9.5 oz.	30
M-Medium	—	—	450-600	300	8 oz.	10 oz.	30
S-Soft	—	—	300-450	200	no limit	no limit	30

[a]For use in exterior walls below grade and for unprotected exterior walls above grade that may be exposed to frost action.

[b]For general use above grade in walls not subjected to frost action or where protected from the weather with two coats of portland cement paint or other satisfactory waterproofing treatment approved by the purchaser.

[c]Regardless of the grade of unit used, protective coatings such as portland cement paint may be desirable on exterior walls for waterproofing purposes. In this connection purchasers should be guided by local experience and the manufacturer's recommendations.

[d]Units with 75 percent or more net area. The classification is based on strength.

[e]Brick intended for use where exposed to temperatures below freezing in the presence of moisture.

[f]Brick intended for use as back-up or interior masonry.

ALLOWABLE SLUMP FOR VARIOUS CONCRETE APPLICATIONS

Types of Construction	Slump (in inches)	
	Maximum	Minimum
Reinforced foundation walls and footings	4	2
Unreinforced footings, caissons, and substructure walls	3	1
Reinforced slabs, beams and walls	5	2
Building columns	5	3
Bridge decks	3	2
Pavements	2	1
Sidewalks, driveways and slabs on ground	4	2
Heavy mass construction	2	1

RECOMMENDED CONCRETE TEMPERATURE FOR COLD WEATHER CONSTRUCTION

Condition of Placement and Curing		Thin Sections	Moderate Sections	Mass Sections
Minimum temperature fresh concrete as mixed for weather indicated, degrees Fahrenheit	Above 30°F	60	55	50
	0 to 30°F	65	60	55
	Below 0°F	70	65	60
Minimum temperature fresh concrete as placed, degrees Fahrenheit		55	50	45
Maximum allowable gradual drop in temperature throughout first 24 hours after end of protection, degrees Fahrenheit		50	40	30

RECOMMENDED DURATION OF PROTECTION FOR CONCRETE PLACED IN COLD WEATHER

Degree of Exposure to Freeze-Thaw	Normal Concrete*	High-Early-Strength Concrete**
No exposure	2 days	1 day
Any exposure	3 days	2 days

*Made with Type I, II, or normal cement.
**Made with Type III or high-early-strength cement, or an accelerator, or an extra 100 lb. of cement.

CURING METHODS

Method	Advantage	Disadvantage
Sprinkling with water or covering with wet burlap	Excellent results if constantly kept wet.	Likelihood of drying between sprinklings. Difficult on vertical walls.
Straw	Insulator in winter.	Can dry out, blow away or burn.
Curing compounds	Easy to apply; inexpensive.	Sprayer needed; inadequate coverage allows drying out; film can be broken or tracked off before curing is completed; unless pigmented, can allow concrete to get too hot.
Moist earth	Cheap, but messy.	Stains concrete, can dry out, removal problem.
Waterproof paper	Excellent protection, prevents drying.	Heavy cost can be excessive. Must be kept in rolls; storage and handling problem.
Plastic film	Absolutely watertight, excellent protection. Light and easy to handle.	Should be pigmented for heat protection. Requires reasonable care and tears must be patched; must be weighed down to prevent blowing away.

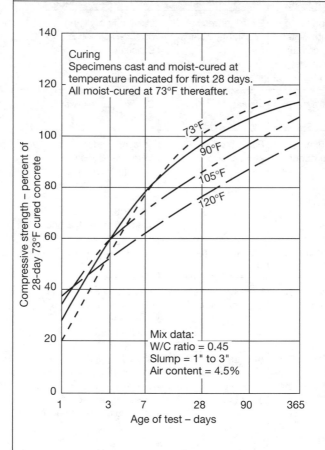

**RELATIVE FINAL STRENGTH OF CONCRETE
AT VARIOUS TEMPERATURES**

Curing
Specimens cast and moist-cured at
temperature indicated for first 28 days.
All moist-cured at 73°F thereafter.

73°F
90°F
105°F
120°F

Compressive strength – percent of
28-day 73°F cured concrete

Mix data:
W/C ratio = 0.45
Slump = 1" to 3"
Air content = 4.5%

Age of test – days

3-24

WATER CONTENT AT VARIOUS TEMPERATURE CHANGES

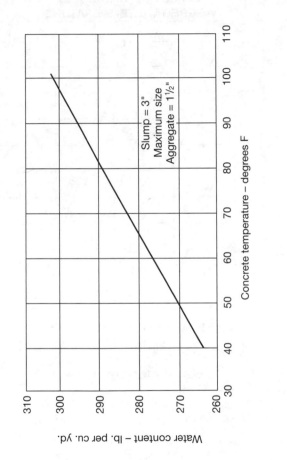

Slump = 3"
Maximum size
Aggregate = 1½"

Concrete temperature – degrees F

Water content – lb. per cu. yd.

EFFECT OF WATER AND AGGREGATE TEMPERATURE

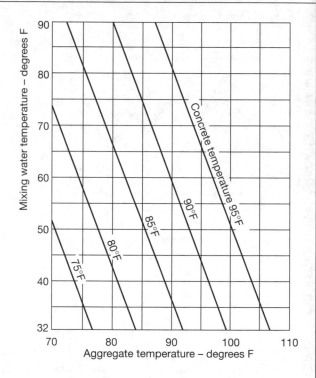

Chart based on the following mix proportions:

Aggregate	3,000 lb.
Moisture in aggregate	60 lb.
Adding mixing water	240 lb.
Cement, at 150°F	564 lb.

CROSS-SECTION END VIEW OF A SIMPLE BLOCK WALL

8" x 8" x 16" concrete block

Footing below frost line

Grade

Same thickness of wall

8"

Twice thickness of wall

1' 4"

Footing for 8" Walls

Cap block

$\frac{1}{2}$" diameter reinforcing bars at 4 ft. centers if wall is more than 4 ft. high

Fill core spaces around bar with concrete

Ground line

Top 4' 0" unreinforced

Lower 2' 0" reinforced

6' 0" max. wall height

18" min. depth

1' 4"

8"

Cross-Section of Garden Wall

Vertical reinforcement rods are placed in the hollow cores at various intervals.

3-27

STANDARD SIZES AND SHAPES OF CONCRETE BLOCKS

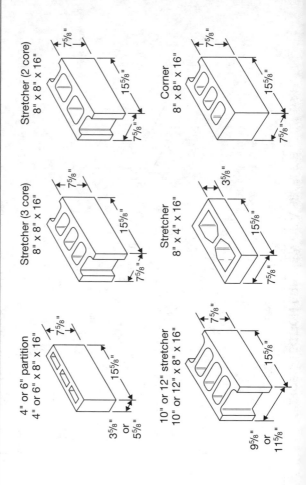

Stretcher (2 core)
8" x 8" x 16"
7⅝"
15⅝"
7⅝"

Corner
8" x 8" x 16"
7⅝"
15⅝"
7⅝"

Stretcher (3 core)
8" x 8" x 16"
7⅝"
15⅝"
7⅝"

Stretcher
8" x 4" x 16"
3⅝"
15⅝"
7⅝"

4" or 6" partition
4" or 6" x 8" x 16"
7⅝"
15⅝"
3⅝" or 5⅝"

10" or 12" stretcher
10" or 12" x 8" x 16"
7⅝"
15⅝"
9⅝" or 11⅝"

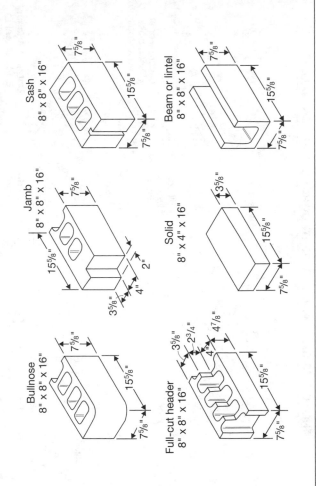

Sash
8" x 8" x 16"
7⅝"
15⅝"
7⅝"

Beam or lintel
8" x 8" x 16"
7⅝"
15⅝"
7⅝"

Jamb
8" x 8" x 16"
7⅝"
15⅝"
2"
4"
3⅝"

Solid
8" x 4" x 16"
3⅝"
15⅝"
7⅝"

Bullnose
8" x 8" x 16"
7⅝"
15⅝"
7⅝"

Full-cut header
8" x 8" x 16"
3⅝"
2¾"
4"
4⅞"
4"
15⅝"
7⅝"

ANCHORING SILLS AND PLATES TO CONCRETE BLOCK WALLS

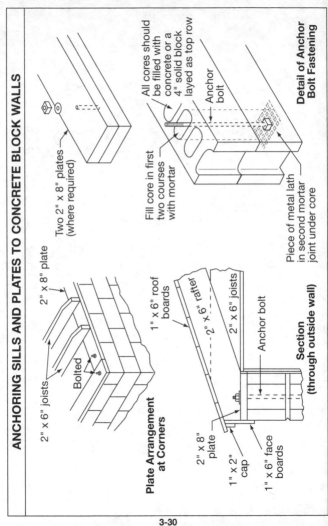

Two 2" x 8" plates (where required)

All cores should be filled with concrete or a 4" solid block layed as top row

Anchor bolt

Fill core in first two courses with mortar

Piece of metal lath in second mortar joint under core

Detail of Anchor Bolt Fastening

2" x 8" plate

2" x 6" joists

Bolted

Plate Arrangement at Corners

1" x 6" roof boards

2" x 6" rafter

2" x 6" joists

Anchor bolt

Section (through outside wall)

2" x 8" plate

1" x 2" cap

1" x 6" face boards

INSTALLATION OF ELECTRICAL SWITCHES AND OUTLET BOXES IN CONCRETE BLOCK WALLS

Type of wiring as per code requirements

Set box in mortar

Cut hole in block with chisel to accommodate switch or box

ALLOWABLE HEIGHT AND MINIMUM NOMINAL THICKNESS FOR CONCRETE MASONRY BEARING WALLS

For buildings up to three stories in height with walls of hollow or solid concrete masonry units. For buildings over three stories, see Building Code. (thickness in inches)

Wall	Residential			Nonresidential		
	one-story	two-story	three-story	one-story	two-story	three-story
3rd story	—	—	8	—	—	12 / 8^d
2nd story	—	8	8	—	12 / 8^d	12
1st story	8 / 6^a	8	8	12 / 8^d	12	12 / 16^c
Basement or foundation		12 or 8^b		12 or 8^b	12	12 or 16

^{a}May be 6" for one-story single-family dwellings and one-story private garages when not more than 9 ft. in height, with an allowance of 6 ft. additional for any gable.

^{b}When foundation wall does not extend more than 4 ft. into the ground the wall may be 8" thick. With special approval of the building official, this depth may be extended to 7 ft. where soil conditions warrant such an extension. In no case shall the total height of 8" or 10" concrete masonry walls, including the foundation wall, exceed 35 ft.

^{c}Must be at least 16" if the total height of the first, second and third stories above the foundation wall, or from a girder or other intermediate supports, is more than 35 ft.

^{d}Top story may be 8" when not more than 12 ft. in height and roof beams are horizontal and total height of masonry wall is not more than 35 ft.

Wall	Cavity Residential			Cavity Nonresidential		
	one-story	two-story	three-story	one-story	two-story	three-story
3rd story	—	—	10	—	—	12 10^f
2nd story	—	10	10	—	12 10^f	12
1st story	10	10	12 or 10^e	12 10^f	12	12^g
Basement or foundation	12, 10 or 8^h (not cavity)			12 or 10 (not cavity)		

[a] In no case shall total height of a 10" cavity wall above the foundation wall exceed 25ft.

[f] Top story may be 10" when not more than 12 ft. in height and roof beams are horizontal and total weight of wall is not more than 35 ft.

[g] In no case shall total height of a cavity wall exceed 35 ft. above the foundation wall, regardless of thickness.

[h] May be 8" for 1½-story single-family dwellings having a maximum height, including the gable, of not over 20 ft. and having nominal 10" cavity walls. Such 8" foundation walls shall be corbelled to provide a bearing the full thickness of the wall above. Total projection not to exceed 2" with top course a full header course not higher than the bottom of the floor joists. Individual projections in corbelling shall not be more than one-third the height of the unit.

INSTALLING VENTILATING AND HEATING DUCTS IN CONCRETE BLOCK WALLS

Ventilator or heating ducts

Partition blocks

CUBIC YARDS OF CONCRETE IN SLABS

Thickness in Inches

Area in square feet (length x width)	4	5	6	8	12
50	0.62	0.77	0.93	1.2	1.9
100	1.2	1.5	1.9	2.5	3.7
200	2.5	3.1	3.7	4.9	7.4
300	3.7	4.6	5.6	7.4	11.1
400	5.0	6.2	7.4	9.9	14.8
500	6.2	7.7	9.3	12.4	18.5

PIGMENTS FOR COLORED CONCRETE FLOOR FINISH

Color Desired	Commercial Names of Colors for Use with Cement	Approximate Quantities Required— lb. per Bag of Cement	
		Light Shade	Medium Shade
Grays, blue-black and black	Germantown lampblack* or carbon black* or black oxide of manganese* or mineral black	$\frac{1}{2}$ $\frac{1}{2}$ 1 1	1 1 2 2
Blue	Ultramarine blue	5	9
Brownish red to dull brick red	Red oxide of iron	5	9
Bright red to vermilion	Mineral turkey red	5	9
Red sandstone to purplish red	Indian red	5	9
Brown to reddish-brown	Metallic brown (oxide)	5	9
Buff, colonial tint and yellow	Yellow ochre or yellow oxide	5 2	9 4
Green	Chromium oxide or greenish blue ultramarine	5 6	9

*Only first-quality lampblack should be used. Carbon black is light in weight and requires thorough mixing. Black oxide or mineral black is most advantageous for general use. For black, use 11 lbs. of oxide for each bag of cement.

LINTEL SIZES

Wall Thickness	3 Feet Steel Angles	3 Feet Wood	4 Feet* Steel Angles	5 Feet* Steel Angles	6 Feet* Steel Angles	7 Feet* Steel Angles	8 Feet* Steel Angles
8"	Two 3 x 3 x ¼	2 x 8 / Two 2 x 4	Two 3 x 3 x ¼	Two 3 x 3 x ¼	Two 3½ x 3½ x ¼	Two 3½ x 3½ x ¼	Two 3½ x 3½ x ¼
12"	Two 3 x 3 x ¼	2 x 12 / Two 2 x 6	Two 3 x 3 x ¼	Two 3½ x 3½ x ¼	Two 3½ x 3½ x ¼	Two 4 x 4 x ¼	Two 4 x 4 x ¼

*Wood lintels should not be used for spans over 3 ft. since they burn out in case of fire and allow the brick to fall.

3-37

NOMINAL BRICK SIZES

Unit Designation	Dimensions			Modular Coursing
	Thickness	Height	Length	
Standard modular	4"	2⅔"	8"	3C = 8"
Engineer	4"	3⅕"	8"	5C = 16"
Economy	4"	4"	8"	1C = 4"
Double	4"	5⅓"	8"	3C = 16"
Roman	4"	2"	12"	2C = 4"
Norman	4"	2⅔"	12"	3C = 8"
Norwegian	4"	3⅕"	12"	5C = 16"
Utility	4"	4"	12"	1C = 4"
Triple	4"	5⅓"	12"	3C = 16"
SCR brick	6"	2⅔"	12"	3C = 8"
6" Norwegian	6"	3⅕"	12"	5C = 16"
6" Jumbo	6"	4"	12"	1C = 4"
8" Jumbo	8"	4"	12"	1C = 4"

HOW VARIOUS BRICKS ARE USED IN OVERLAP BONDING

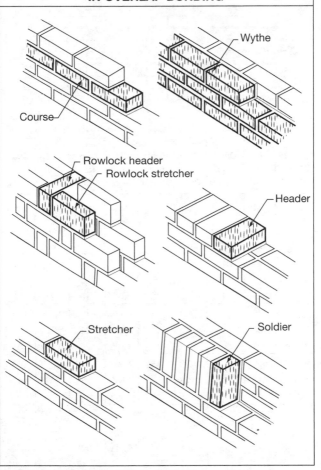

Wythe

Course

Rowlock header
Rowlock stretcher

Header

Stretcher

Soldier

AMOUNT OF BRICK AND MORTAR NEEDED FOR VARIOUS WALL SIZES

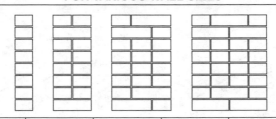

Area of Wall (sq. ft.)	4-Inch Wall		8-Inch Wall		12-Inch Wall		16-Inch Wall	
	No. of Brick	Cubic ft. of Mortar	No. of Brick	Cubic ft. of Mortar	No. of Brick	Cubic ft. of Mortar	No. of Brick	Cubic ft. of Mortar
1	6.2	0.075	12.4	0.195	18.5	0.314	24.7	0.433
10	62	1	124	2	185	$3\frac{1}{2}$	247	$4\frac{1}{2}$
20	124	2	247	4	370	$6\frac{1}{2}$	493	9
30	185	$2\frac{1}{2}$	370	6	555	$9\frac{1}{2}$	740	13
40	247	$3\frac{1}{2}$	493	8	740	13	986	$17\frac{1}{2}$
50	309	4	617	10	925	16	1,233	22
60	370	5	740	12	1,109	19	1,479	26
70	432	$5\frac{1}{2}$	863	14	1,294	22	1,725	31
80	493	$6\frac{1}{2}$	986	16	1,479	25	1,972	35
90	555	7	1,109	18	1,664	28	2,218	39
100	617	8	1,233	20	1,849	32	2,465	44
200	1,233	15	2,465	39	3,697	63	4,929	87
300	1,849	23	3,697	59	5,545	94	7,393	130
400	2,465	30	4,929	78	7,393	126	9,857	173
500	3,081	38	6,161	98	9,241	157	12,321	217

AMOUNT OF BRICK AND MORTAR NEEDED FOR VARIOUS WALL SIZES *(cont.)*

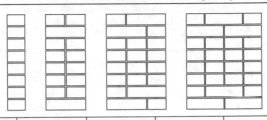

Area of Wall (sq. ft.)	4-Inch Wall		8-Inch Wall		12-Inch Wall		16-Inch Wall	
	No. of Brick	Cubic ft. of Mortar	No. of Brick	Cubic ft. of Mortar	No. of Brick	Cubic ft. of Mortar	No. of Brick	Cubic ft. of Mortar
600	3,697	46	7,393	117	11,089	189	14,786	260
700	4,313	53	8,625	137	12,937	220	17,250	303
800	4,929	61	9,857	156	14,786	251	19,714	347
900	5,545	68	11,089	175	16,634	283	22,178	390
1,000	6,161	76	12,321	195	18,482	314	24,642	433
2,000	12,321	151	24,642	390	36,963	628	49,284	866
3,000	18,482	227	36,963	584	55,444	942	73,926	1,299
4,000	24,642	302	49,284	779	73,926	1,255	98,567	1,732
5,000	30,803	377	61,605	973	92,407	1,568	123,209	2,165
6,000	36,963	453	73,926	1,168	110,888	1,883	147,851	2,599
7,000	43,124	528	86,247	1,363	129,370	2,197	172,493	3,032
8,000	49,284	604	98,567	1,557	147,851	2,511	197,124	3,465
9,000	55,444	679	110,888	1,752	166,332	2,825	221,776	3,898
10,000	61,605	755	123,209	1,947	184,813	3,139	246,418	4,331

Note: Mortar joints are $1/2$" thick.

NUMBER OF CONCRETE BLOCKS BY LENGTH OF WALL

Length of Wall	No. of Units	Length of Wall	No. of Units	Length of Wall	No. of Units	Length of Wall	No. of Units	Length of Wall	No. of Units	Length of Wall	No. of Units
0' 8"	$\frac{1}{2}$	20' 8"	$15\frac{1}{2}$	40' 8"	$30\frac{1}{2}$	60' 8"	$45\frac{1}{2}$	80' 8"	$60\frac{1}{2}$	100' 8"	$75\frac{1}{2}$
1' 4"	1	21' 4"	16	41' 4"	31	61' 4"	46	81' 4"	61	101' 4"	76
2' 0"	$1\frac{1}{2}$	22' 0"	$16\frac{1}{2}$	42' 0"	$31\frac{1}{2}$	62' 0"	$46\frac{1}{2}$	82' 0"	$61\frac{1}{2}$	102' 0"	$76\frac{1}{2}$
2' 8"	2	22' 8"	17	42' 8"	32	62' 8"	47	82' 8"	62	102' 8"	77
3' 4"	$2\frac{1}{2}$	23' 4"	$17\frac{1}{2}$	43' 4"	$32\frac{1}{2}$	63' 4"	$47\frac{1}{2}$	83' 4"	$62\frac{1}{2}$	103' 4"	$77\frac{1}{2}$
4' 0"	3	24' 0"	18	44' 0"	33	64' 0"	48	84' 0"	63	104' 0"	78
4' 8"	$3\frac{1}{2}$	24' 8"	$18\frac{1}{2}$	44' 8"	$33\frac{1}{2}$	64' 8"	$48\frac{1}{2}$	84' 8"	$63\frac{1}{2}$	104' 8"	$78\frac{1}{2}$
5' 4"	4	25' 4"	19	45' 4"	34	65' 4"	49	85' 4"	64	105' 4"	79
6' 0"	$4\frac{1}{2}$	26' 0"	$19\frac{1}{2}$	46' 0"	$34\frac{1}{2}$	66' 0"	$49\frac{1}{2}$	86' 0"	$64\frac{1}{2}$	106' 0"	$79\frac{1}{2}$
6' 8"	5	26' 8"	20	46' 8"	35	66' 8"	50	86' 8"	65	106' 8"	80
7' 4"	$5\frac{1}{2}$	27' 4"	$20\frac{1}{2}$	47' 4"	$35\frac{1}{2}$	67' 4"	$50\frac{1}{2}$	87' 4"	$65\frac{1}{2}$	107' 4"	$80\frac{1}{2}$
8' 0"	6	28' 0"	21	48' 0"	36	68' 0"	51	88' 0"	66	108' 0"	81
8' 8"	$6\frac{1}{2}$	28' 8"	$21\frac{1}{2}$	48' 8"	$36\frac{1}{2}$	68' 8"	$51\frac{1}{2}$	88' 8"	$66\frac{1}{2}$	108' 8"	$81\frac{1}{2}$
9' 4"	7	29' 4"	22	49' 4"	37	69' 4"	52	89' 4"	67	109' 4"	82
10' 0"	$7\frac{1}{2}$	30' 0"	$22\frac{1}{2}$	50' 0"	$37\frac{1}{2}$	70' 0"	$52\frac{1}{2}$	90' 0"	$67\frac{1}{2}$	110' 0"	$82\frac{1}{2}$

Courses	Length	Courses	Length	Courses	Length	Courses	Length	Courses	Length	Courses	Length
8	10' 8"	23	30' 8"	38	50' 8"	53	70' 8"	68	90' 8"	83	110' 8"
8½	11' 4"	23½	31' 4"	38½	51' 4"	53½	71' 4"	68½	91' 4"	83½	111' 4"
9	12' 0"	24	32' 0"	39	52' 0"	54	72' 0"	69	92' 0"	84	112' 0"
9½	12' 8"	24½	32' 8"	39½	52' 8"	54½	72' 8"	69½	92' 8"	84½	112' 8"
10	13' 4"	25	33' 4"	40	53' 4"	55	73' 4"	70	93' 4"	85	113' 4"
10½	14' 0"	25½	34' 0"	40½	54' 0"	55½	74' 0"	70½	94' 0"	85½	114' 0"
11	14' 8"	26	34' 8"	41	54' 8"	56	74' 8"	71	94' 8"	86	114' 8"
11½	15' 4"	26½	35' 4"	41½	55' 4"	56½	75' 4"	71½	95' 4"	86½	115' 4"
12	16' 0"	27	36' 0"	42	56' 0"	57	76' 0"	72	96' 0"	87	116' 0"
12½	16' 8"	27½	36' 8"	42½	56' 8"	57½	76' 8"	72½	96' 8"	87½	116' 8"
13	17' 4"	28	37' 4"	43	57' 4"	58	77' 4"	73	97' 4"	88	117' 4"
13½	18' 0"	28½	38' 0"	43½	58' 0"	58½	78' 0"	73½	98' 0"	88½	118' 0"
14	18' 8"	29	38' 8"	44	58' 8"	59	78' 8"	74	98' 8"	89	118' 8"
14½	19' 4"	29½	39' 4"	44½	59' 4"	59½	79' 4"	74½	99' 4"	89½	119' 4"
15	20' 0"	30	40' 0"	45	60' 0"	60	80' 0"	75	100' 0"	90	120' 0"

Note: Based on units 15⅝" long and half units 7⅝" long with ⅜" thick head joints.

MATERIAL QUANTITIES PER CUBIC FOOT OF MORTAR

| | Quantities by Volume | | | |
| | Mortar Type and Proportions by Volume | | | |
Material	M 1:¼:3	S 1:½:4½	N 1:1:6	O 1:2:9
Cement	0.333	0.222	0.167	0.111
Lime	0.083	0.111	0.167	0.222
Sand	1.000	1.000	1.000	1.000

| | Quantities by Weight | | | |
| | Mortar Type and Proportions by Volume | | | |
Material	M 1:¼:3	S 1:½:4½	N 1:1:6	O 1:2:9
Cement	31.33	20.89	15.67	10.44
Lime	3.33	4.44	6.67	8.89
Sand	80.00	80.00	80.00	80.00

NUMBER OF COURSES OF CONCRETE BLOCKS BY HEIGHT OF WALL

Height of Wall	No. of Units	Height of Wall	No. of Units	Height of Wall	No. of Units	Height of Wall	No. of Units
0' 8"	1	8' 8"	13	16' 8"	25	24' 8"	37
1' 4"	2	9' 4"	14	17' 4"	26	25' 4"	38
2' 0"	3	10' 0"	15	18' 0"	27	26' 0"	39
2' 8"	4	10' 8"	16	18' 8"	28	26' 8"	40
3' 4"	5	11' 4"	17	19' 4"	29	27' 4"	41
4' 0"	6	12' 0"	18	20' 0"	30	28' 0"	42
4' 8"	7	12' 8"	19	20' 8"	31	28' 8"	43
5' 4"	8	13' 4"	20	21' 4"	32	29' 4"	44
6' 0"	9	14' 0"	21	22' 0"	33	30' 0"	45
6' 8"	10	14' 8"	22	22' 8"	34	30' 8"	46
7' 4"	11	15' 4"	23	23' 4"	35	31' 4"	47
8' 0"	12	16' 0"	24	24' 0"	36	32' 0"	48

Note: Based on units 7⅝" high and ⅜" mortar joints.

3-45

MORTAR JOINTS

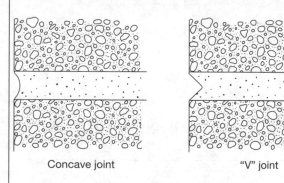

Concave joint   "V" joint

For Exterior and Interior Walls

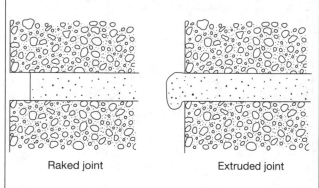

Raked joint   Extruded joint

For Interior Walls

Note: Raked and extruded joints are not recommended for exterior walls in cold climates.

RATIO OF CEMENT, LIME AND SAND FOR VARIOUS MORTAR MIXES

Mix by Volume, Cement-Lime-Sand	Cement (sacks)	Lime (lb.)	Sand (cu. yd.)
1 – 0.05 – 2	13.00	26	0.96
1 – 0.05 – 3	9.00	18	1.00
1 – 0.05 – 4	6.75	44	1.00
1 – 0.10 – 2	13.00	52	.96
1 – 0.10 – 3	9.00	36	1.00
1 – 0.10 – 4	6.75	27	1.00
1 – 0.25 – 2	12.70	127	.94
1 – 0.25 – 3	9.00	90	1.00
1 – 0.25 – 4	6.75	67	1.00
1 – 0.50 – 2	12.40	250	.92
1 – 0.50 – 3	8.80	175	.98
1 – 0.50 – 4	6.75	135	1.00
1 – 0.50 – 5	5.40	110	1.00
1 – 1 – 3	8.60	345	.95
1 – 1 – 4	6.60	270	.98
1 – 1 – 5	5.40	210	1.00
1 – 1 – 6	4.50	180	1.00
1 – 1.5 – 3	8.10	485	.90
1 – 1.5 – 4	6.35	380	.94
1 – 1.5 – 5	5.30	320	.98
1 – 1.5 – 6	4.50	270	1.00
1 – 1.5 – 7	3.85	230	1.00
1 – 1.5 – 8	3.40	205	1.00
1 – 2 – 4	6.10	490	.90
1 – 2 – 5	5.10	410	.94
1 – 2 – 6	4.40	350	.98
1 – 2 – 7	3.85	310	1.00
1 – 2 – 8	3.40	270	1.00
1 – 2 – 9	3.00	240	1.00

RECOMMENDED MORTAR MIXES

Proportions by Volume

Type of Service	Cement	Hydrated Lime	Mortar Sand in Damp, Loose Condition
For ordinary service	1 masonry cement or 1 portland cement	— 1 to 1/4	2 to 3 4 to 6
Subject to extremely heavy loads, violent winds, earthquakes or severe frost action. Isolated piers.	1 masonry cement plus 1 portland cement or 1 portland cement	— 0 to 1/4	4 to 6 2 to 3

CUBIC FEET OF MORTAR FOR COLLAR JOINTS

Cubic Feet of Mortar per 100 sq. ft. of Wall

1/4-in. joint	3/8-in. joint	1/2-in. joint
2.08	3.13	4.17

Note: Cubic feet per 1000 units = $\dfrac{10 \times \text{cubic feet per 100 sq. ft. of wall}}{\text{number of units per square foot of wall}}$

3-48

NOMINAL MODULAR SIZES OF STRUCTURAL LOAD-BEARING TILE

Backup Tile

	Face Dimension in Wall	
Thickness (inches)	Height (inches)	Length (inches)
4	$2\frac{2}{3}$	8 or 12
4	$5\frac{1}{3}$	12^1
4	8	8 or 12
4	$10\frac{2}{3}$	12
6	$5\frac{1}{3}$	12
6	8	12^1
6	$10\frac{2}{3}$	12
8	$5\frac{1}{3}$	12
8	8	8 or 12^1
8	$10\frac{2}{3}$	12

Wall Tile

Thickness (inches)	Height (inches)	Length (inches)
4	$5\frac{1}{3}$	12
4	8	8 or 12
4	12	12
6	$5\frac{1}{3}$	12
6	12	12
8	$5\frac{1}{3}$	12
8	6	12
8	8	8, 12 or 16
8	12	12
10	8	12
10	12	12
12	12	12

Note: Nominal sizes include the thickness of the standard mortar joint for all dimensions.
[1]Includes header and stretcher units.

MATERIALS USED FOR HOLLOW CLAY TILE WALLS

Wall Thickness — Tile Size — Wall Area (sq. ft.)	4 inches — 4 x 5 x 12		8 inches — 8 x 5 x 12	
	Number of Tile	Mortar (cu. ft.)	Number of Tile	Mortar (cu. ft.)
1	2.1	0.045	2.1	0.09
10	21	.45	21	.9
100	210	4.5	210	9.0
200	420	9.0	420	18
300	630	13.5	630	27
400	840	18.0	840	36
500	1050	22.5	1050	45
600	1260	27.0	1260	54
700	1470	31.5	1470	63
800	1680	36.0	1680	72
900	1890	40.5	1890	81
1000	2100	45.0	2100	90

Note: Quantities are based on 1/2-inch thick mortar joint.

NONMODULAR BRICK AND MORTAR, SINGLE WYTHE WALLS, RUNNING BOND						
	With ³/₈-in. Joints			With ½-in. Joints		
Size of Brick (inches) th. x ht. x lth.	Number of Brick per 100 sq. ft.	Cubic Feet of Mortar per 100 sq. ft.	Cubic Feet of Mortar per 1000 Brick	Number of Brick per 100 sq. ft.	Cubic Feet of Mortar per 100 sq. ft.	Cubic Feet of Mortar per 1000 Brick
2³/₄ x 2³/₄ x 9³/₄	455	3.2	7.1	432	4.5	10.4
2⁵/₈ x 2³/₄ x 8³/₄	504	3.4	6.8	470	4.1	8.7
3³/₄ x 2¹/₄ x 8	655	5.8	8.8	616	7.2	11.7
3³/₄ x 2³/₄ x 8	551	5.0	9.1	522	6.4	12.2

Note: No allowances for breakage or waste.

MATERIALS NEEDED FOR END CONSTRUCTION USING HOLLOW TILE

Wall Thickness	4 Inches		6 Inches		8 Inches		10 Inches	
Tile Size	4 x 12 x 12		6 x 12 x 12		8 x 12 x 12		10 x 12 x 12	
Wall Area (sq. ft.)	Number of Tile	Cu. ft. Mortar	Number of Tile	Cu. ft. Mortar	Number of Tile	Cu. ft. Mortar	Number of Tile	Cu. ft. Mortar
1	0.93	0.025	0.93	0.036	0.93	0.049	0.93	0.06
10	9.3	.25	9.3	.36	9.3	.49	9.3	.4
100	93	2.5	93	3.6	93	4.9	93	6
200	186	5.0	186	7.2	186	9.8	186	12
300	279	7.5	279	10.8	279	14.7	279	18
400	372	10.0	372	14.4	372	19.6	372	24
500	465	12.5	465	18.0	465	24.5	465	30
600	558	15.0	558	21.6	558	29.4	558	36
700	651	17.5	651	25.2	651	34.3	651	42
800	837	22.5	837	32.4	837	44.1	837	54
900	744	20.0	744	28.8	744	39.2	744	48
1000	930	25.0	930	36.0	930	49.0	930	60

Note: Quantities are based on ½-inch thick mortar joint.

MODULAR BRICK AND MORTAR, SINGLE WYTHE WALLS, RUNNING BOND

Nominal Size of Brick (inches) th. x ht. x lth.	Number of Brick per 100 sq. ft.	Cubic Feet of Mortar			
		Per 100 sq. ft.		Per 1000 Brick	
		3/8-in. Joints	1/2-in. Joints	3/8-in. Joints	1/2-in. Joints
4 x 2 2/3 x 8	675	5.5	7.0	8.1	10.3
4 x 3 1/5 x 8	563	4.8	6.1	8.6	10.9
4 x 4 x 8	450	4.2	5.3	9.2	11.7
4 x 5 1/3 x 8	338	3.5	4.4	10.2	12.9
4 x 2 x 12	600	6.5	8.2	10.8	13.7
4 x 2 2/3 x 12	450	5.1	6.5	11.3	14.4
4 x 3 1/5 x 12	375	4.4	5.6	11.7	14.9
4 x 4 x 12	300	3.7	4.8	12.3	15.7
4 x 5 1/3 x 12	225	3.0	3.9	13.4	17.1
6 x 2 2/3 x 12	450	7.9	10.2	17.5	22.6
6 x 3 1/5 x 12	375	6.8	8.8	18.1	23.4
6 x 4 x 12	300	5.6	7.4	19.1	24.7

Note: No allowances for breakage or waste.

Roofing

Rafters

Joists

Ceiling
Plate, two 2" x 4"s

Brick

Sheathing

1" air
space

2" x 4" studs
@ 16" O.C.

Drywall or plaster

Building
paper

**Wall Section —
Brick Veneer
on Frame**

Finish floor

Subfloor

Metal ties

Flashing

Joist

Header

Weep holes
@ 2' 0" O.C.

Finish grade

6" solid unit

Foundation
walls

10"

Metal ties

**Alternate
Foundation
Detail**

Flashing

Weep holes
@ 2' 0" O.C.

Anchor

Brick corbel

2"

8"

3-54

BRICK VENEER

Roofing

Rafters

Ceiling

Plate, two 2" x 4"s

Brick

Sheathing

Metal ties

2" x 4" studs
@ 16" O.C.

1" air
space

**Wall Section —
Brick Veneer
on Frame**

Metal ties

Drywall or plaster

Flashing

Finish floor

Weep holes
@ 2' 0" O.C.

4" concrete
slab with
turned down
footing

Finish grade

Grout

Metal ties

Flashing

Concrete slab

**Alternate
Foundation
Detail**

Weep holes
@ 2' 0" O.C.

Perimeter
insulation

6" tile and 4" brick
foundation wall

3-55

SIZE OF FLUE LININGS

Nominal Size of Chimney	Outside Dimensions of Flue Linings	Weight per Feet	Length of Piece
4" x 8"	4½" x 8½"	14 lbs.	2 feet
4" x 13"	4½" x 13"	20 lbs.	2 feet
8" x 8"	8½" x 8½"	18 lbs.	2 feet
8" x 12"	8½" x 13"	30 lbs.	2 feet
8" x 16"	8½" x 18"	36 lbs.	2 feet
12" x 12"	13" x 13"	38 lbs.	2 feet
12" x 16"	13" x 18"	45 lbs.	2 feet
16" x 16"	18" x 18"	62 lbs.	2 feet

NUMBER OF BRICK FOR CHIMNEYS

per Foot of Height

Size and Number of Flues	Number of Brick	Cubic Feet Mortar
(a) 1-8" x 8" flue	27	0.5
(b) 1-8" x 12" flue	31	0.5
(c) 1-12" x 12" flue	35	0.6
(d) 2-8" x 8" flue	46	0.8
(e) 1-8" x 8" and 1-8" x 12" flue	51	0.9
(f) 2-8" x 12" flue	55	0.10
(g) 2-8" x 12" flue	53	0.9
(h) 1-8" x 12" and 12" x 12" flue	58	1.0
(i) 2-12" x 12" flue	62	1.1
(j) 2-8" x 8" and 1-8" x 12" flue	70	1.2
(k) 1-8" x 12" and 2-12" x 12" flue	83	1.4
(l) 1-8" x 12" and 2-12" x 12" flue	70	1.2
(m) 1-8" x 8" extending 12" from face of wall	18	0.4
(n) 1-8" x 8" extending 8" from face of wall	9	0.3
(o) 1-8" x 8" extending 4" from face of wall	0	0.0

3-57

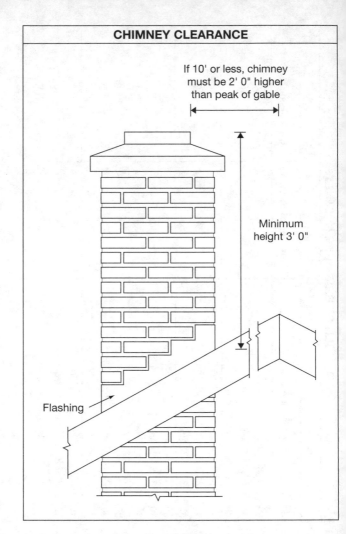

CHIMNEY CLEARANCE

If 10' or less, chimney must be 2' 0" higher than peak of gable

Minimum height 3' 0"

Flashing

3-58

STANDARD SIZES OF MODULAR CLAY FLUE LININGS

Minimum Net Inside Area (sq. in.)	Nominal[1] Dimensions (inches)	Outside[2] Dimensions (inches)	Minimum Wall Thickness (inches)	Approximate Maximum Outside Corner Radius (inches)
15	4 x 8	3.5 x 7.5	0.5	1
20	4 x 12	3.5 x 11.5	0.625	1
27	4 x 16	3.5 x 15.5	0.75	1
35	8 x 8	7.5 x 7.5	0.625	2
57	8 x 12	7.5 x 11.5	0.75	2
74	8 x 16	7.5 x 15.5	0.875	2
87	12 x 12	11.5 x 11.5	0.875	3
120	12 x 16	11.5 x 15.5	1.0	3
162	16 x 16	15.5 x 15.5	1.125	4
208	16 x 20	15.5 x 19.5	1.25	4
262	20 x 20	19.5 x 19.5	1.375	5
320	20 x 24	19.5 x 23.5	1.5	5
385	24 x 24	23.5 x 23.5	1.625	6

[1]Cross section of flue lining shall fit within rectangle of dimension corresponding to nominal size.
[2]Length in each case shall be 24 ± 0.5 in.

CROSS-SECTION DRAWINGS

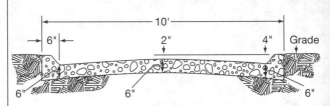

2" Crown

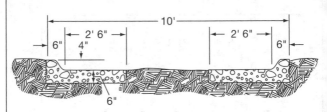

**Parallel Strips
with Curb**

Parallel Strips

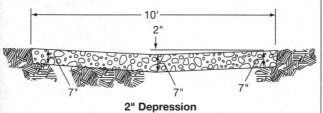

2" Depression

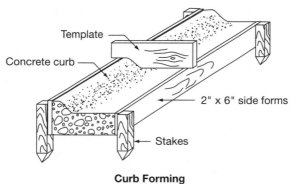

Template

Concrete curb

2" x 6" side forms

Stakes

Curb Forming

GLASS BLOCK QUANTITIES

Plan   Section

No. of Units	$5\frac{3}{4}" \times 5\frac{3}{4}" \times 3\frac{7}{8}"$	$7\frac{3}{4}" \times 7\frac{3}{4}" \times 3\frac{7}{8}"$	$11\frac{3}{4}" \times 11\frac{3}{4}" \times 3\frac{7}{8}"$	No. of Units	$5\frac{3}{4}" \times 5\frac{3}{4}" \times 3\frac{7}{8}"$	$7\frac{3}{4}" \times 7\frac{3}{4}" \times 3\frac{7}{8}"$	$11\frac{3}{4}" \times 11\frac{3}{4}" \times 3\frac{7}{8}"$
1	6"	8"	1' 0"	19	9' 6"	12' 8"	19' 0"
2	1' 0"	1' 4"	2' 0"	20	10' 0"	13' 4"	20' 0"
3	1' 6"	2' 0"	3' 0"	21	10' 6"	14' 0"	21' 0"
4	2' 0"	2' 8"	4' 0"	22	11' 0"	14' 8"	22' 0"
5	2' 6"	3' 4"	5' 0"	23	11' 6"	15' 4"	23' 0"
6	3' 0"	4' 0"	6' 0"	24	12' 0"	16' 0"	24' 0"
7	3' 6"	4' 8"	7' 0"	25	12' 6"	16' 8"	25' 0"
8	4' 0"	5' 4"	8' 0"	26	13' 0"	17' 4"	
9	4' 6"	6' 0"	9' 0"	27	13' 6"	18' 0"	
10	5' 0"	6' 8"	10' 0"	28	14' 0"	18' 8"	
11	5' 6"	7' 4"	11' 0"	29	14' 6"	19' 4"	
12	6' 0"	8' 0"	12' 0"	30	15' 0"	20' 0"	
13	6' 6"	8' 8"	13' 0"	31	15' 6"	20' 8"	
14	7' 0"	9' 4"	14' 0"	32	16' 0"	21' 4"	
15	7' 6"	10' 0"	15' 0"	33	16' 6"	22' 0"	
16	8' 0"	10' 8"	16' 0"	34	17' 0"	22' 8"	
17	8' 6"	11' 4"	17' 0"				
18	9' 0"	12' 0"	18' 0"				

3-62

CONCRETE SLABS, WALKS AND DRIVEWAYS

Expansion joint

Strike board

Divider form board

Wooden tamper

2" x 4" or 2" x 6" side forms

Subgrade (see note)

Note: Subgrade may consist of cinder, gravel or other suitable material where conditions require. The subgrade should be well-tamped before placing concrete.

3-63

GRADIENTS

Desirable Grades	Maximum, %	Minimum, %
Streets (concrete)	8	0.50
Parking (concrete)	5	0.50
Service areas (concrete)	5	0.50
Main approach walls to buildings	4	1
Stoops or entries to buildings	2	1
Collector walks	8	1
Terraces and sitting areas	2	1
Grass recreational areas	3	2
Mowed banks of grass	3:1 slope	—
Unmowed banks	2:1 slope	—

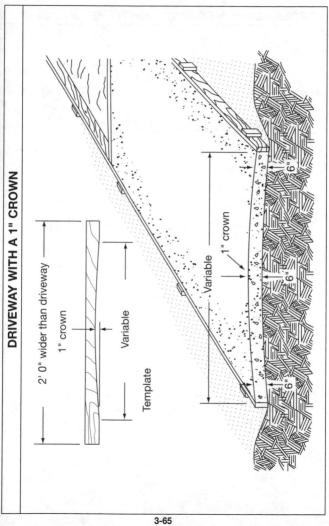

DRIVEWAY WITH A 1" CROWN

2' 0" wider than driveway

1" crown

Variable

Template

Variable

1" crown

6"

6"

6"

CROSS-SECTION OF SINGLE-CAR LEVEL DRIVEWAY

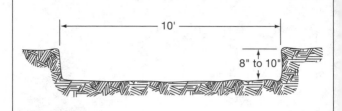

10'

8" to 10"

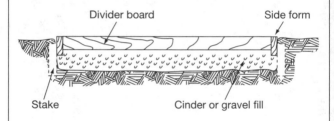

Divider board

Side form

Stake

Cinder or gravel fill

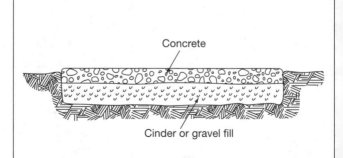

Concrete

Cinder or gravel fill

APPROXIMATE WEIGHT AND STRENGTH OF STONE

Kind of Stone	Crushing Weight lbs. (per cu. ft.)	Shearing Strength lbs. (per sq. in.)	Strength lbs. (per sq. in.)
Sandstone	150	8,000	1,500
Granite	170	15,000	2,000
Limestone	170	6,000	1,000
Marble	170	10,000	1,400
Slate	175	15,000	—
Trap Rock	185	20,000	—

MORTAR FOR GLASS BLOCK

For One Cubic Foot of Mortar	Mix by Volume	
	1:¼:3	1:1:6
Portland cement	0.3 bag	0.16 bag
Hydrated lime	0.06 bag	0.13 bag
Plastering sand	0.92 cubic foot	1.0 cubic foot
Waterproofing	0.3 quart	0.2 quart

FOUNDATION AND WALL THICKNESSES

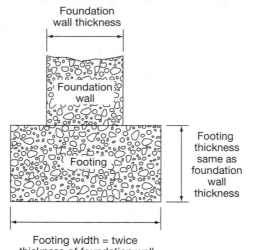

Foundation wall thickness

Foundation wall

Footing

Footing thickness same as foundation wall thickness

Footing width = twice thickness of foundation wall

Approximate Ratio of Foundation Size to the Wall it Supports

REINFORCING BAR NUMBERS AND DIMENSIONS

Bar Sizes*		Weight (lbs. per. ft.)	Cross-Sectional Area (sq. in.)
Old (inches)	New Numbers		
1/4	2	0.166	0.05
3/8	3	0.376	0.1105
1/2	4	0.668	0.1963
5/8	5	1.043	0.3068
3/4	6	1.502	0.4418
7/8	7	2.044	0.6013
1	8	2.670	0.7854
1 / square	9	3.400	1.0000
1 1/8 square	10	4.303	1.2656
1 1/4 square	11	5.313	1.5625

Note: Bar numbers are based on the nearest number of 1/8 inch included in the nominal diameter of the bar. Bars numbered 9, 10 and 11 are round bars and equivalent in weight and nominal cross-sectional area to the old type 1", 1 1/8" and 1 1/4" square bars.

HOOK DIMENSIONS BENDING OF REINFORCING BARS

Recommended sizes – 180° hook

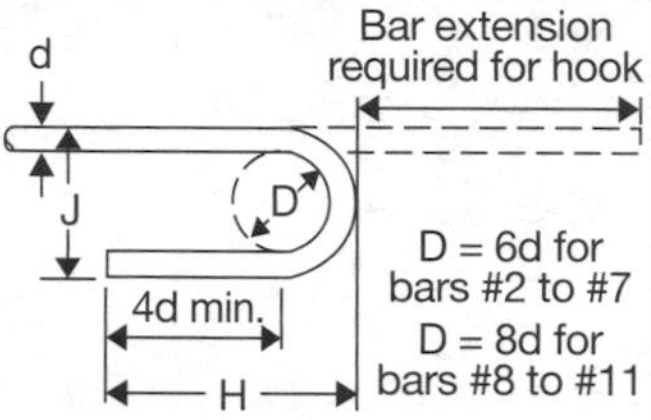

Bar extension required for hook

d, J, D, 4d min., H

D = 6d for bars #2 to #7
D = 8d for bars #8 to #11

J	Bar exten.	Approx. H	Bar size D
2	4	3½	#2
3	5	4	3
4	6	4½	4
5	7	5	5
6	8	6	6
7	10	7	7
10	13	9	8
11¼	15	10¼	9
12½	17	11¼	10
14	19	12¾	11

Minimum sizes – 180° hook

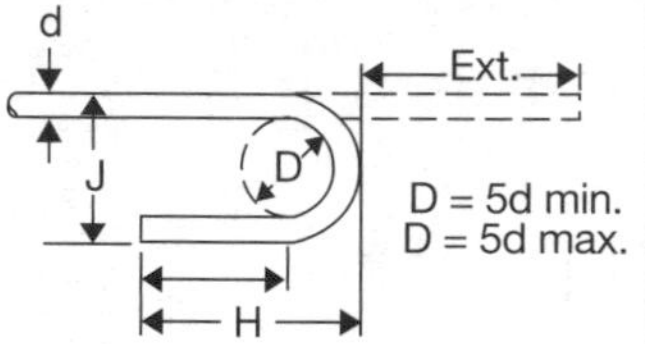

d, Ext., J, D

D = 5d min.
D = 5d max.

H

Note: Minimum size hooks to be used only for special conditions; do not use for hard-grade steel.

J	Bar exten.	Approx. H	Bar size D
1¾	4	3½	#2
2¾	5	4	3
3½	5	4¼	4
4¼	6	4¾	5
5¼	7	5¾	6
6	9	6½	7
7	10	7½	8
8	11	8½	9
9	13	9½	10
10	14	10½	11

Recommended sizes – 90° hook

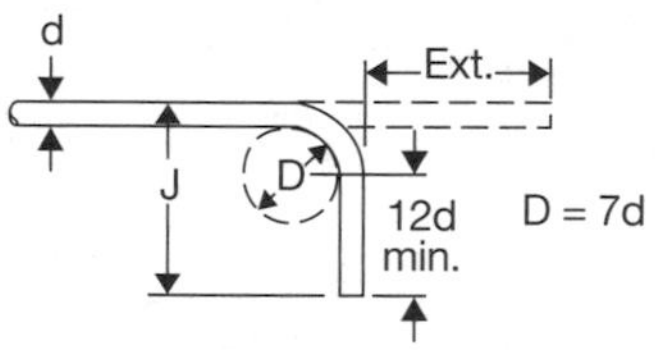

d, Ext., J, D

12d min.

D = 7d

Bar exten.	Approx. J	Bar size D
3	3½	#2
3	4	3
3	4½	4
4	5	5
4	6	6
5	7	7
6	9	8
7	10	9
8	11¼	10
9	12½	11

Recommended sizes – 135° stirrup hook

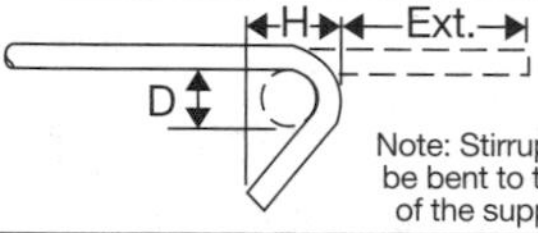

H, Ext., D

D = 5d

Note: Stirrup hooks may be bent to the diameter of the supporting bars.

Bar exten.	Approx. H	Bar size D
3½	2	#2
4	2¼	3
4½	2½	4
5	2¾	5

CHAPTER 4
Carpentry

FASTENER APPLICATIONS		
Nails — Common Applications		
Joining	**Size & Type**	**Placement**
Wall Framing		
Top plate	8d common	
	16d common	
Header	8d common	
	16d common	
Header to joist	16d common	
Studs	8d common	
	16d common	
Wall Sheathing		
Boards	8d common	6" o.c.
Plywood ($^5/_{16}$", $^3/_8$", $^1/_2$")	6d common	6" o.c.
Plywood ($^5/_8$", $^3/_4$")	8d common	6" o.c.
Fiberboard	1$^3/_4$" galv roofing nail	6" o.c.
	8d galv common nail	6" o.c.
Foamboard	Cap nail length sufficient for penetration of $^1/_2$" into framing	12" o.c.
Gypsum	1$^3/_4$" galv roofing nail	6" o.c.
	8d galv common nail	6" o.c.
Subflooring	8d common	10"-12" o.c.
Underlayment	1$^1/_4$" x 14 ga annular underlayment nail	6" o.c. edges 12" o.c. face
Roof Framing		
Rafters, beveled or notched	12d common	
Rafter to joist	16d common	
Joist to rafter and stud	10d common	
Ridge beam	8d & 16d common	
Roof Sheathing		
Boards	8d common	
Plywood ($^5/_{16}$", $^3/_8$", $^1/_2$")	6d common	12" o.c. and 6" o.c. edges
Plywood ($^5/_8$", $^3/_4$")	8d common	12" o.c. and 6" o.c. edges

FASTENER APPLICATIONS (cont.)

Joining	Size & Type	Placement
Roofing, Asphalt		
New construction shingles and felt	⅞" through 1½" galv roofing	4 per shingle
Re-roofing application shingles and felt	1¾" or 2" galv roofing	4 per shingle
Roof deck/insulation	Thickness of insulation plus 1" insulation roof deck nail	
Roofing, Wood Shingles		
New construction	3d-4d galv shingle	2-3 per shingle
Re-roofing application	5d-6d galv shingle	2-3 per shingle
Soffit	6d-8d galv common	12" o.c. max.
Siding		
Bevel and lap Drop and shiplap Plywood	Aluminum nails are recommended for optimum performance	Consult siding manufacturer's application instructions
Hardboard	Galvanized hardboard siding nail Galvanized box nail	Consult siding manufacturer's application instructions
Doors, Windows, Moldings, Furring		
Wood strip to masonry Wood strip to stud or joist	Nail length is determined by thickness of siding and sheathing. Nails should penetrate at least 1½" into solid wood framing	
Paneling		
Wood	4d-8d casing-finishing	24" o.c.
Hardboard	2" x 16 ga annular	8" o.c.
Plywood	3d casing-finishing	8" o.c.
Gypsum	1¼" annular drywall	6" o.c.
Lathing	4d common glued	4" o.c.
Exterior Projects		
Decks, patios, etc.	8d-16d hot dipped galvanized common	

4-2

DRYWALL SCREWS

Description	No.	Length	Applications
Bugle Phillips 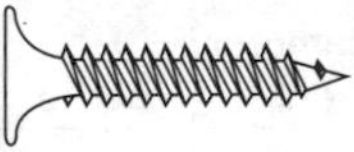	1E 2E 3E 4E 5R 6R 7R 8R	6 x 1 6 x $1\frac{1}{8}$ 6 x $1\frac{1}{4}$ 6 x $1\frac{5}{8}$ 6 x 2 6 x $2\frac{1}{4}$ 8 x $2\frac{1}{2}$ 8 x 3	For attaching drywall to metal studs from 25 ga through 20 ga
Coarse Thread 	1C 2C 3C 4C 5C 6C	6 x 1 6 x $1\frac{1}{8}$ 6 x $1\frac{1}{4}$ 6 x $1\frac{5}{8}$ 6 x 2 6 x $2\frac{1}{4}$	For attaching drywall to 25 ga metal studs and attaching drywall to wood studs
Pan Framing 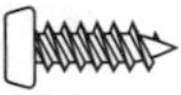	19	6 x $\frac{7}{16}$	For attaching stud to track up to 20 ga
HWH Framing 	21 22 35	6 x $\frac{7}{16}$ 8 x $\frac{9}{16}$ 10 x $\frac{3}{4}$	For attaching stud to track up to 20 ga where hex head is desired
K-Lath 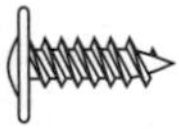	28	8 x $\frac{9}{16}$	For attaching wire lath K-lath to 20 ga studs
Laminating 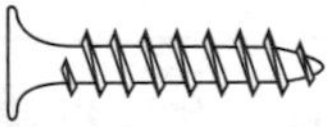	8	10 x $1\frac{1}{2}$	Type G laminating screw for attaching gypsum to gypsum; a temporary fastener
Trim head 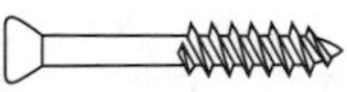	9 10	6 x $1\frac{5}{8}$ 6 x $2\frac{1}{4}$	Trim head screw for attaching wood trim and base to 25 ga studs

NAIL SIZES AND APPROXIMATE NUMBER PER POUND

Size	Length (in.)	Common Diameter (in.)	Common No. per lb.	Box Diameter (in.)	Box No. per lb.
4d	$1\frac{1}{2}$	.102	316	.083	473
5d	$1\frac{3}{4}$	.102	271	.083	406
6d	2	.115	181	.102	236
7d	$2\frac{1}{4}$	.115	161	.102	210
8d	$2\frac{1}{2}$	.131	106	.115	145
10d	3	.148	69	.127	94
12d	$3\frac{1}{4}$	.148	63	.127	88
16d	$3\frac{1}{2}$	.165	49	.134	71
20d	4	.203	31	.148	52
30d	$4\frac{1}{2}$	.220	24	.148	46
40d	5	.238	18	.165	35

LUMBER PRODUCT CLASSIFICATION

	Thickness (in.)	Width (in.)
Board lumber	1"	2" or more
Light framing	2" to 4"	2" to 4"
Studs	2" to 4"	2" to 4" 10' and shorter
Structural light framing	2" to 4"	2" to 4"
Joists and planks	2" to 4"	6" and wider
Beams and stringers	5" and thicker	more than 2" greater than thickness
Posts and timbers	5" x 5" and larger	not more than 2" greater than thickness
Decking	2" to 4"	4" to 12" wide
Siding	thickness expressed by dimension of butt edge	
Moldings	size at thickest and widest points	

Note: Lengths of lumber generally are 6 feet and longer
in multiples of 2'.

Size of Timber (in inches)	10'	12'	14'	16'	18'	20'	22'	24'
1 x 2	$1\frac{2}{3}$	2	$2\frac{1}{3}$	$2\frac{2}{3}$	3	$3\frac{1}{3}$	$3\frac{2}{3}$	4
1 x 3	$2\frac{1}{2}$	3	$3\frac{1}{2}$	4	$4\frac{1}{2}$	5	$5\frac{1}{2}$	6
1 x 4	$3\frac{1}{3}$	4	$4\frac{2}{3}$	$5\frac{1}{3}$	6	$6\frac{2}{3}$	$7\frac{1}{3}$	8
1 x 5	$4\frac{1}{6}$	5	$5\frac{5}{6}$	$6\frac{2}{3}$	$7\frac{1}{2}$	$8\frac{1}{3}$	$9\frac{1}{6}$	10
1 x 6	5	6	7	8	9	10	11	12
1 x 8	$6\frac{2}{3}$	8	$9\frac{1}{3}$	$10\frac{2}{3}$	12	$13\frac{1}{3}$	$14\frac{2}{3}$	16
1 x 10	$8\frac{1}{3}$	10	$11\frac{2}{3}$	$13\frac{1}{3}$	15	$16\frac{2}{3}$	$18\frac{1}{3}$	20
1 x 12	10	12	14	16	18	20	22	24
1 x 14	$11\frac{2}{3}$	14	$16\frac{1}{3}$	$18\frac{2}{3}$	21	$23\frac{1}{3}$	$25\frac{2}{3}$	28
1 x 16	$13\frac{1}{3}$	16	$18\frac{2}{3}$	$21\frac{1}{3}$	24	$26\frac{2}{3}$	$29\frac{1}{3}$	32
1 x 20	$16\frac{2}{3}$	20	$23\frac{1}{3}$	$26\frac{2}{3}$	30	$33\frac{1}{3}$	$36\frac{2}{3}$	40
1¼ x 4	$4\frac{1}{6}$	5	$5\frac{5}{6}$	$6\frac{2}{3}$	$7\frac{1}{2}$	$8\frac{1}{3}$	$9\frac{1}{6}$	10
1¼ x 6	$6\frac{1}{4}$	$7\frac{1}{2}$	$8\frac{3}{4}$	10	$11\frac{1}{4}$	$12\frac{1}{2}$	$13\frac{3}{4}$	15
1¼ x 8	$8\frac{1}{3}$	10	$11\frac{2}{3}$	$13\frac{1}{3}$	15	$16\frac{2}{3}$	$18\frac{1}{3}$	20
1¼ x 10	$10\frac{1}{3}$	$12\frac{1}{2}$	$14\frac{1}{2}$	$16\frac{2}{3}$	$18\frac{2}{3}$	$20\frac{5}{6}$	$22\frac{5}{6}$	25
1¼ x 12	$12\frac{1}{2}$	15	$17\frac{1}{2}$	20	$22\frac{1}{2}$	25	$27\frac{1}{2}$	30
1½ x 4	5	6	7	8	9	10	11	12
1½ x 6	$7\frac{1}{2}$	9	$10\frac{1}{2}$	12	$13\frac{1}{2}$	15	$16\frac{1}{2}$	18
1½ x 8	10	12	14	16	18	20	22	24
1½ x 10	$12\frac{1}{2}$	15	$17\frac{1}{2}$	20	$22\frac{1}{2}$	25	$27\frac{1}{2}$	30

	15	18	21	24	27	30	33	36
1½ x 12	15	18	21	24	27	30	33	36
2 x 4	$6\frac{2}{3}$	8	$9\frac{1}{3}$	$10\frac{2}{3}$	12	$13\frac{1}{3}$	$14\frac{2}{3}$	16
2 x 6	10	12	14	16	18	20	22	24
2 x 8	$13\frac{1}{3}$	16	$18\frac{2}{3}$	$21\frac{1}{3}$	24	$26\frac{2}{3}$	$29\frac{1}{3}$	32
2 x 10	$16\frac{2}{3}$	20	$23\frac{1}{3}$	$26\frac{2}{3}$	30	$33\frac{1}{3}$	$36\frac{2}{3}$	40
2 x 12	20	24	28	32	36	40	44	48
2 x 14	$23\frac{1}{3}$	28	$32\frac{2}{3}$	$37\frac{1}{3}$	42	$46\frac{2}{3}$	$51\frac{1}{3}$	56
2 x 16	$26\frac{2}{3}$	32	$37\frac{1}{3}$	$42\frac{2}{3}$	48	$53\frac{1}{3}$	$58\frac{2}{3}$	64
2½ x 12	25	30	35	40	45	50	55	60
2½ x 14	$29\frac{1}{6}$	35	$40\frac{5}{6}$	$46\frac{2}{3}$	$52\frac{1}{2}$	$58\frac{1}{3}$	$64\frac{1}{6}$	70
2½ x 16	$33\frac{1}{3}$	40	$46\frac{2}{3}$	$53\frac{1}{3}$	60	$66\frac{2}{3}$	$73\frac{1}{3}$	80
3 x 6	15	18	21	24	27	30	33	36
3 x 8	20	24	28	32	36	40	44	48
3 x 10	25	30	35	40	45	50	55	60
3 x 12	30	36	42	48	54	60	66	72
3 x 14	35	42	49	56	63	70	77	84
3 x 16	40	48	56	64	72	80	88	96
4 x 4	$13\frac{1}{3}$	16	$18\frac{2}{3}$	$21\frac{1}{3}$	24	$26\frac{2}{3}$	$29\frac{1}{3}$	32
4 x 6	20	24	28	32	36	40	44	48
4 x 8	$26\frac{2}{3}$	32	$37\frac{1}{3}$	$42\frac{2}{3}$	48	$53\frac{1}{3}$	$58\frac{2}{3}$	64
4 x 10	$33\frac{1}{3}$	40	$46\frac{2}{3}$	$53\frac{1}{3}$	60	$66\frac{2}{3}$	$73\frac{1}{3}$	80
4 x 12	40	48	56	64	72	80	88	96
4 x 14	$46\frac{2}{3}$	56	$65\frac{1}{3}$	$74\frac{2}{3}$	84	$93\frac{1}{3}$	$102\frac{2}{3}$	112

CONVERTING LINEAR FEET TO BOARD FEET (cont.)

1 x 2: 1/6 x length
1 x 3: 1/4 x length
1 x 4: 1/3 x length
1 x 6: 1/2 x length
1 x 8: 2/3 x length
1 x 10: 5/6 x length
1 x 12: 1 x length
1 x 14: 1 1/16 x length
1 x 16: 1 1/3 x length

2 x 4: 2/3 x length
2 x 6: 1 x length
2 x 8: 1 1/3 x length
2 x 10: 1 2/3 x length
2 x 12: 2 x length
4 x 4: 1 1/3 x length
4 x 6: 2 x length
4 x 8: 2 2/3 x length

4-8

BOARD FOOTAGE

The unit of measure for lumber is the board foot. This is a piece 1 in. thick and 12 in. square or its equivalent (144 cu. in.).

A board 1 x 12 and 10 ft. long will contain 10 bd. ft. If it were only 6 in. wide, it would be 5 bd. ft. If the original board had been 2 in. thick, it would have contained 20 bd. ft.

The following formula can be applied to any size piece where the total length is given in feet:

$$\text{Bd. ft.} = \frac{\text{No. pcs.} \times T \times W \times L}{12}$$

To find the number of board feet in six pieces of lumber that measure 1" x 8" x 14':

$$\text{Bd. ft.} = \frac{6 \times 1 \times \overset{4}{8} \times 14}{\underset{\underset{1}{2}}{12}} = 56$$

$$= 56 \text{ bd. ft.}$$

Stock less than 1 in. thick is figured as though it were 1 in. When the stock is thicker than 1 in., the nominal size is used. When this size contains a fraction such as 1¼, change it to an improper fraction (⁵⁄₄) and place the numerator above the formula line and the denominator below.

To find the board footage in two pieces of lumber that measure 1¼" x 10" x 8':

$$\text{Bd. ft.} = \frac{\overset{1}{2} \times \overset{5}{5} \times 10 \times \overset{2}{8}}{\underset{\underset{1}{2}}{4} \quad \underset{3}{12} \quad 3} = 50$$

$$= 16⅔ \text{ bd ft.}$$

Use the nominal size of the material when figuring the footage.

SOFTWOOD LUMBER CLASSIFICATIONS AND GRADES

Appearance Grades

Boards		
Selects	B & better C Select D Select	(IWP-Supreme) (IWP-Choice) (IWP-Quality)
Finish	Superior Prime E	
Paneling	Clear (any select or finish grade) No. 2 common selected for knotty paneling No. 3 common selected for knotty paneling	
Siding (bevel, bungalow)	Superior Prime	

		Alternate Board Grades
Boards Sheathing	No. 1 Common (IWP-Colonial) No. 2 Common (IWP-Sterling) No. 3 Common (IWP-Standard) No. 4 Common (IWP-Utility)	Select Merchantable Construction Standard Utility

Note: Names of grades and their specifications will vary.

Specification Check List

☐ Grades listed in order of quality.

☐ Include all species suited to project.

☐ For economy, specify lowest grade that will satisfy job requirement.

☐ Specify surface texture desired.

☐ Specify moisture content suited to project.

☐ Specify $\overset{W}{\underset{WP}{\bigcirc}}$ grade stamp. For finish and exposed pieces, specify stamp on back or ends.

Western Red Cedar

Finish Paneling and Ceiling	Clear Heart A B
Bevel Siding	Clear — V. G. Heart A — Bevel Siding B — Bevel Siding C — Bevel Siding

Dimension		
Light Framing 2" to 4" thick 2" to 4" wide	Construction Standard Utility Economy	This category for use where high strength values are NOT required; such as studs, plates, sills, cripples, blocking, etc.
	Stud Economy Stud	An optional all-purpose grade limited to 10 feet and shorter. Characteristics affecting strength and stiffness values are limited so that the "Stud" grade is suitable for all stud uses, including load bearing walls.
Structural Light Framing 2" to 4" thick 2" to 4" wide	Select Structural No. 1 No. 2 No. 3 Economy	These grades are designed to fit those engineering applications where higher bending strength ratios are needed in light framing sizes. Typical uses would be for trusses, concrete pier wall forms, etc.
Structural Joists and Planks 2" to 4" thick 6" and wider	Select structural No. 1 No. 2 No. 3 Economy	These grades are designed especially to fit in engineering applications for lumber six inches and wider, such as joists, rafters and general framing uses.

Timbers			
Beams and Stringers	Select Structural No. 1 No. 2 (No. 1 mining) No. 3 (No. 2 mining)	**Posts and Timbers**	Select Structural No. 1 No. 2 (No. 1 mining) No. 3 (No. 2 mining)

STANDARD LUMBER SIZES/NOMINAL, DRESSED

Product	Description	Nominal Size		Dressed Dimensions		
				Thicknesses and Widths (inches)		
		Thickness (inches)	Width (inches)	Surfaced Dry	Surfaced Unseasoned	Lengths (feet)
Framing	S4S	2	2	1½	1 9/16	6 ft. and longer
		3	3	2½	2 9/16	in multiples
		4	4	3½	3 9/16	of 1'
			6	5½	5⅝	
			8	7¼	7½	
			10	9¼	9½	
			12	11¼	11½	
			Over 12	Off ¾	Off ½	
				Thickness (inches)	Width (inches)	
Timbers	Rough or S4S	5 and larger		½ off nominal		Same
		Nominal Size		Dressed Dimensions Surfaced Dry		
		Thickness (inches)	Width (inches)	Thickness (inches)	Width (inches)	Lengths (feet)
	2" single T&G	2	6	1 ½	5	6 ft. and longer
Decking			8		6¾	in multiples
			10		8¾	of 1'
Decking is usually surfaced to single T&G in 2" thickness and double T&G in 3" and 4" thicknesses			12		10¾	
	3" and 4" double T&G	3	6	2½	5¼	
		4		3½		

	Nominal Thickness	Nominal Width	Surfaced Size	Length
Flooring (D & M), (S2S & CM)	3/8 1/2 5/8 1 1¼ 1½	2 3 4 5 6	5/16 7/16 9/16 3/4 1 1¼	4 ft. and longer in multiples of 1'
Ceiling and Partition (S2S & CM)	3/8 1/2 5/8 3/4	3 4 5 6	5/16 7/16 9/16 11/16	4 ft. and longer in multiples of 1'
Factory and Shop Lumber S2S	1 (4/4) 1¼ (5/4) 1½ (6/4) 1¾ (7/4) 2 (8/4) 2½ (10/4) 3 (12/4) 4 (16/4)	5 and wider (4" and wider in 4/4 No. 1 Shop and 4/4 No. 2 Shop)	25/32 (4/4) 1 5/32 (5/4) 1 19/32 (6/4) 1 19/32 (7/4) 1 13/16 (8/4) 2 3/8 (10/4) 2 3/4 (12/4) 3 3/4 (16/4)	Usually sold random width / 4 ft. and longer in multiples of 1'

Abbreviations

Abbreviated descriptions appearing in the size table are explained below.

S1S — Surfaced one side.	S1S2E — Surfaced one side, two edges.
S2S — Surfaced two sides.	CM — Center matched.
S4S — Surfaced four sides.	D & M — Dressed and matched.
S1S1E — Surfaced one side, one edge.	
T & G — Tongue and grooved.	
EV1S — Edge vee on one side.	
S1E — Surfaced one edge.	

NAILING SCHEDULE

Connection	Nailing[1]
Joist to sill or girder, toenail	3-8d
Bridging to joist, toenail each end	2-8d
1" x 6" subfloor or less to each joist, face nail	2-8d
Wider than 1" x 6" subfloor to each joist, face nail	3-8d
2" subfloor to joist or girder, blind and face nail	2-16d
Sole plate to joist or blocking, face nail	16d at 16" o.c.
Top plate to stud, end nail	2-16d
Stud to sole plate	4-8d, toenail or 2-16d, end nail
Double studs, face nail	16d at 24" o.c.
Doubled top plates, face nail	16d at 16" o.c.
Top plates, laps and intersections, face nail	2-16d
Continuous header, two pieces	16d at 16" o.c. along each edge
Ceiling joists to plate, toenail	3-8d
Continuous header to stud, toenail	4-8d
Ceiling joists, laps over partitions, face nail	3-16d
Ceiling joists to parallel rafters, face nail	3-16d
Rafter to plate, toenail	3-8d
1" brace to each stud and plate, face nail	2-8d
1" x 8" sheathing or less to each bearing, face nail	2-8d
Wider than 1" x 8" sheathing to each bearing, face nail	3-8d
Built-up corner studs	16d at 24" o.c.
Built-up girder and beams	20d at 32" o.c. at top and bottom and staggered 2-20d at ends and at each splice
2" planks	2-16d at each bearing

NAILING SCHEDULE (cont.)

Connection	Nailing[1]
Plywood and particleboard:[5]	
Subfloor, roof and wall sheathing (to framing):	
$\frac{1}{2}$" and less	6d[2]
$\frac{19}{32}$" – $\frac{3}{4}$"	8d[3] or 6d[4]
$\frac{7}{8}$" – 1"	8d[2]
$1\frac{1}{8}$" – $1\frac{1}{4}$"	10d[3] or 8d[4]
Combination Subfloor-underlayment (to framing):	
$\frac{3}{4}$" and less	6d[4]
$\frac{7}{8}$" – 1"	8d[4]
$1\frac{1}{8}$" – $1\frac{1}{4}$"	10d[3] or 8d[4]
Panel Siding (to framing):	
$\frac{1}{2}$" or less	6d[6]
$\frac{5}{8}$"	8d[6]
Fiberboard Sheathing:[7]	
$\frac{1}{2}$"	No. 11 ga[8]
	6d[3]
	No. 16 ga[9]
$\frac{25}{32}$"	No. 11 ga[8]
	8d[3]
	No. 16 ga[9]

[1]Common or box nails may be used except where otherwise stated.

[2]Common or deformed shank.

[3]Common.

[4]Deformed shank.

[5]Nails spaced at 6 inches on center at edges, 12 inches at intermediate supports except 6 inches at all supports where spans are 48 inches or more. Nails for wall sheathing may be common, box or casing.

[6]Corrosion-resistant siding or casing nails.

[7]Fasteners spaced 3 inches on center at exterior edges and 6 inches on center at intermediate supports.

[8]Corrosion-resistant roofing nails with $\frac{7}{16}$-inch diameter head and $1\frac{1}{2}$-inch length for $\frac{1}{2}$-inch sheathing and $1\frac{3}{4}$-inch length for $\frac{25}{32}$-inch sheathing.

[9]Corrosion-resistant staples with nominal $\frac{7}{16}$-inch crown and $1\frac{1}{8}$-inch length for $\frac{1}{2}$-inch sheathing and $1\frac{1}{2}$-inch length for $\frac{25}{32}$-inch sheathing.

SIZE, HEIGHT AND SPACING OF WOOD STUDS

Stud Size (inches)	Bearing Walls				Nonbearing Walls	
	Laterally Unsupported Stud Heights[1] (feet)	Supporting Roof and Ceiling Only	Supporting One Floor, Roof and Ceiling	Supporting Two Floors, Roof and Ceiling	Laterally Unsupported Stud Height[1] (feet)	Spacing (inches)
		Spacing (inches)				
2 x 3[2]	—	—	—	—	10	16
2 x 4	10	24	16	—	14	24
3 x 4	10	24	24	16	14	24
2 x 5	10	24	24	—	16	24
2 x 6	10	24	24	16	20	24

Utility grade studs shall be spaced no more than 16 inches on center, or support more than a roof and ceiling, or exceed 8 feet in height for exterior walls and load-bearing walls or 10 feet for interior nonbearing walls.

[1] Listed heights are distances between points of lateral support placed perpendicular to the plane of the wall. Increases in unsupported height are permitted where justified by an analysis.

[2] Shall not be used in exterior walls.

4-16

NUMBER AND SPACING OF WOOD JOISTS FOR ANY FLOOR

Length of Span	Spacing of Joists									
	12"	16"	20"	24"	30"	36"	42"	48"	54"	60"
6	7	6	5	4	3	3	3	3	2	2
7	8	6	5	5	4	4	3	3	3	2
8	9	7	6	5	4	4	3	3	3	3
9	10	8	6	6	5	4	4	3	3	3
10	11	9	7	6	5	4	4	4	3	3
11	12	9	8	7	5	5	4	4	3	3
12	13	10	8	7	6	5	4	4	4	3
13	14	11	9	8	6	5	5	4	4	4
14	15	12	9	8	7	6	5	5	4	4
15	16	12	10	9	7	6	5	5	4	4
16	17	13	11	9	7	6	6	5	5	4
17	18	14	11	10	8	7	6	5	5	4
18	19	15	12	10	8	7	6	6	5	4
19	20	15	12	11	9	7	6	6	5	5
20	21	16	13	11	9	8	7	6	5	5
21	22	17	14	12	9	8	7	6	6	5
22	23	18	14	12	10	8	7	7	6	5
23	24	18	15	13	10	9	8	7	6	6
24	25	19	15	13	11	9	8	7	6	6
25	26	20	16	14	11	9	8	7	7	6
26	27	21	17	14	11	10	8	8	7	6
27	28	21	17	15	12	10	9	8	7	6
28	29	22	18	15	12	10	9	8	7	7
29	30	23	18	16	13	11	9	8	7	7
30	31	24	19	16	13	11	10	9	8	7
31	32	24	20	17	13	11	10	9	8	7
32	33	25	20	17	14	12	10	9	8	7
33	34	26	21	18	14	12	10	9	8	8
34	35	27	21	18	15	12	11	10	9	8
35	36	27	22	19	15	13	11	10	9	8
36	37	28	23	19	15	13	11	10	9	8
37	38	29	23	20	16	13	12	10	9	8
38	39	30	24	20	16	14	12	11	9	9
39	40	30	24	21	17	14	12	11	10	9
40	41	31	25	21	17	14	12	11	10	9

One joist has been added to each of the above quantities to take care of extra joist required at end of span. Add for doubling joists under all partitions.

BUILT-UP WOOD HEADER DOUBLE OR TRIPLE ON TWO 4" x 4" POSTS

Span (in feet)	Weight (in pounds) safely supported by:							
	2 - 2 x 6	2 - 2 x 8	2 - 2 x 10	2 - 2 x 12	3 - 2 x 6	3 - 2 x 8	3 - 2 x 10	3 - 2 x 12
4	2250	4688	5000	5980	3780	5850	7410	8970
6	1680	3126	5000	5980	2520	4689	7410	8970
8	—	2657	3761	5511	—	3985	5641	8266
10	—	2125	3008	4409	—	3187	4512	6613
12	—	—	2507	3674	—	—	3760	5511
14	—	—	—	3149	—	—	—	4723

	ALLOWABLE SPANS FOR HEADERS UNDER DIFFERENT LOAD CONDITIONS						
	Outside Walls				Inside Walls		
Nominal depth of header (inches)	Roof with or without attic storage	Roof with or without attic storage + one floor	Roof with or without attic storage + two floors	Little attic storage	Full attic storage, or roof load, or little attic storage + one floor	Full attic storage + one floor, or roof load + one floor, or little attic storage + two floors	Full attic storage + two floors, or roof load + two floors
4	4'	2'	2'	4'	2'	No	No
6	6'	5'	4'	6'	3'	2' 6"	2'
8	8'	7'	6'	8'	4'	3'	3'
10	10'	8'	7'	10'	5'	4'	3' 6"
12	12'	9'	8'	12' 6"	6'	5'	4'

STEEL PLATE HEADER ON TWO 4" x 4" POSTS

Weight (in pounds) safely supported by wood sides and plate

Span (in feet)	2 – 2 x 8 + 7½" by			2 – 2 x 10 + 9½" by			2 – 2 x 12 + 11½" by		
Plate	3/8"	7/16"	1/2"	3/8"	7/16"	1/2"	3/8"	7/16"	1/2"
10	6,754	7,538	8,242	10,973	12,199	13,418	15,933	17,729	19,604
12	5,585	6,216	6,827	9,095	10,131	11,106	13,224	14,517	16,265
14	4,756	5,293	5,811	7,751	8,623	9,463	11,295	12,561	13,876
16	—	4,481	5,036	6,746	7,494	8,221	9,815	10,953	12,086
18	—	—	—	5,942	6,606	7,158	8,675	9,652	10,647
20	—	—	—	—	—	6,466	7,746	8,618	9,408

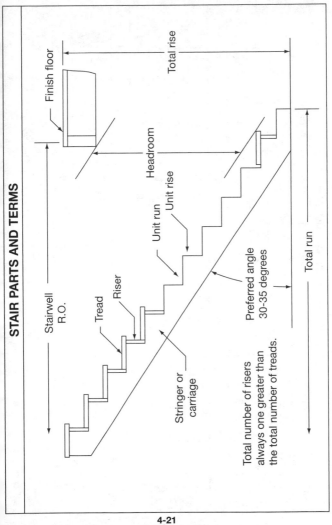

STAIR PARTS AND TERMS

Finish floor

Total rise

Stairwell R.O.

Headroom

Unit rise

Unit run

Tread

Riser

Stringer or carriage

Preferred angle 30-35 degrees

Total run

Total number of risers always one greater than the total number of treads.

4-21

DIMENSIONS FOR STRAIGHT STAIRS

Height Floor-to-Floor H	Number of Risers	Height of Risers R	Width of Treads T	Total Run L	Minimum Headroom Y	Well Opening U
8' 0"	12	8"	9"	8' 3"	6' 6"	8' 1"
8' 0"	13	7 3/8" +	9 1/2"	9' 6"	6' 6"	9' 2 1/2"
8' 0"	13	7 3/8" +	10"	10' 0"	6' 6"	9' 8 1/2"
8' 6"	13	7 7/8" -	9"	9' 0"	6' 6"	8' 3"
8' 6"	14	7 5/16" -	9 1/2"	10' 3 1/2"	6' 6"	9' 4"
8' 6"	14	7 5/16" -	10"	10' 10"	6' 6"	9' 10"
9' 0"	14	7 11/16" +	9"	9' 9"	6' 6"	8' 5"
9' 0"	15	7 3/16" +	9 1/2"	11' 1"	6' 6"	9' 6 1/2"
9' 0"	15	7 3/16" +	10"	11' 8"	6' 6"	9' 11 1/2"
9' 6"	15	7 5/8" -	9"	10' 6"	6' 6"	8' 6 1/2"
9' 6"	16	7 1/8"	9 1/2"	11' 10 1/2"	6' 6"	9' 7"
9' 6"	16	7 1/8"	10"	12' 6"	6' 6"	10' 1"

Dimensions shown under well opening "U" are based on 6' 6" minimum headroom. If headroom is increased, well opening also increases.

DIMENSIONS FOR STAIRS WITH LANDINGS

Height Floor-to-Floor H	Number of Risers	Height of Risers R	Width of Tread T	Run		Run		Run	
				Number of Risers	L	Number of Risers	L2		
8' 0"	13	$7\frac{3}{8}$" +	10"	11	8' 4" + W	2	0' 10" + W		
8' 6"	14	$7\frac{5}{16}$" −	10"	12	9' 2" + W	2	0' 10" + W		
9' 0"	15	$7\frac{3}{16}$" +	10"	13	10' 0" + W	2	0' 10" + W		
9' 6"	16	$7\frac{1}{8}$"	10"	14	10' 10" + W	2	0' 10" + W		

STAIR RISERS, TREADS AND DIMENSIONS

Total Rise Floor-to-Floor H	Number of Risers	Height of Riser R	Number of Treads	Width of Run T	Total Run L	Well Opening U	Length of Carriage	Use Stock Tread Width	Dimension of Nosing Projection
8' 0"	12	8"	11	9 1/2"	8' 8 1/2"	9' 1"	11' 4 5/8"	10 1/2"	1"
	14	6 7/8"	13	10 5/8"	11' 6 1/8"	10' 10"	13' 8 1/2"	11 1/2"	7/8"
8' 4"	13	7 11/16"	12	9 13/16"	9' 9 3/4"	10' 0"	12' 5 1/2"	10 1/2"	11/16"
	14	7 1/8"	13	10 3/8"	11' 2 7/8"	11' 0"	13' 7 5/8"	11 1/2"	1 1/8"
8' 6"	13	7 7/8"	12	9 5/8"	9' 7 1/2"	9' 2"	12' 5 1/4"	10 1/2"	7/8"
	14	7 5/16"	13	10 3/16"	11' 1/2"	10' 8"	13' 7"	11 1/2"	1 5/16"
8' 9"	14	7 1/2"	13	9 1/4"	10' 1/4"	9' 5"	12' 10 3/4"	10 1/2"	1 1/4"
	14	7 1/2"	13	10"	10' 10"	10' 1"	13' 6 1/2"	11 1/2"	1 1/2"
	14	7 5/8"	13	9 3/8"	10' 1 7/8"	9' 5"	13' 1 1/4"	10 1/2"	1 1/8"
8' 11"	14	7 5/8"	13	9 1/16"	9' 9 7/8"	9' 0"	12' 10"	10 1/2"	1 7/16"
	14	7 5/8"	13	10 1/4"	11' 1 1/4"	10' 2"	13' 10 1/4"	11 1/2"	1 1/4"
9' 1"	14	7 13/16"	13	9 11/16"	10' 6"	9' 5"	13' 5 3/4"	10 1/2"	13/16"
	15	7 1/4"	14	10 1/4"	11' 11 1/2"	10' 8"	14' 7 3/4"	11 1/2"	1 1/4"

Well openings based on minimum head height of 6' 8". Dimensions based on 2" x 10" floor joist.

STANDARD DOOR OPENING SIZES

Width of Opening	1¾-Inch Thick Doors Height of Opening					1⅜-Inch Thick Doors Height of Opening	
2' 0"	6' 8"	7' 0"	7' 2"	7' 10"	8' 0"	6' 8"	7' 0"
2' 4"	6' 8"	7' 0"	7' 2"	7' 10"	8' 0"	6' 8"	7' 0"
2' 6"	6' 8"	7' 0"	7' 2"	7' 10"	8' 0"	6' 8"	7' 0"
2' 8"	6' 8"	7' 0"	7' 2"	7' 10"	8' 0"	6' 8"	7' 0"
3' 0"	6' 8"	7' 0"	7' 2"	7' 10"	8' 0"	6' 8"	7' 0"
3' 4"	6' 8"	7' 0"	7' 2"	7' 10"	8' 0"	—	—
3' 6"	6' 8"	7' 0"	7' 2"	7' 10"	8' 0"	—	—
3' 8"	6' 8"	7' 0"	7' 2"	7' 10"	8' 0"	—	—
4' 0"	6' 8"	7' 0"	7' 2"	7' 10"	8' 0"	—	—

ALLOWABLE CLEAR SPANS FOR VARIOUS SIZES AND SPECIES OF JOISTS

Ceiling Joists

Species	Grade	Span (feet and inches)					
		2 x 4		2 x 6		2 x 8	
		16" oc	24" oc	16" oc	24" oc	16" oc	24" oc
Douglas Fir-Larch	2 & better	11-6	10-0	18-1	15-7	23-10	20-7
	3	9-9	7-11	14-8	11-11	19-4	15-9
Douglas Fir South	2 & better	10-6	9-2	16-6	14-5	21-9	19-0
	3	9-5	7-9	14-2	11-7	18-9	15-3
Hem-Fir	2 & better	10-9	9-5	16-11	13-11	22-4	18-4
	3	8-7	7-0	13-1	10-8	17-2	14-0
Mountain Hemlock	2 & better	9-11	8-8	15-7	13-8	20-7	18-0
	3	8-11	7-3	13-3	10-10	17-6	14-4
Mountain Hemlock-Hem-Fir	2 & better	9-11	8-8	15-7	13-8	20-7	18-0
	3	8-7	7-0	13-1	10-8	17-2	14-0
Western Hemlock	2 & better	10-9	9-5	16-11	14-6	22-4	19-1
	3	9-0	7-5	13-9	11-3	18-1	14-10

Species	Grade						
Engelmann Spruce-Alpine Fir (Engelmann Spruce-Lodgepole Pine)	2 & better	9-11	8-8	15-6	12-8	20-7	16-8
	3	7-10	6-5	11-9	9-7	15-6	12-8
Lodgepole Pine	2 & better	10-3	8-11	16-1	13-3	21-2	17-6
	3	8-4	6-9	12-7	10-3	16-7	13-6
Ponderosa Pine-Sugar Pine (Ponderosa Pine-Lodgepole Pine)	2 & better	9-11	8-8	15-7	12-10	20-7	16-10
	3	8-0	6-6	12-0	9-10	15-10	12-11
White Woods (Western Woods)	2 & better	9-7	8-5	15-1	12-6	20-2	16-5
	3	7-10	6-5	11-9	9-8	15-6	12-8
Idaho White Pine	2 & better	10-3	8-5	15-3	12-6	20-0	16-5
	3	7-10	6-5	11-9	9-8	15-6	12-8
Western Cedars	2 & better	9-7	8-5	15-1	13-2	20-0	17-6
	3	8-3	6-9	12-7	10-3	16-7	13-7

Design Criteria:
Strength—5 lbs. per sq. ft. dead load plus 10 lbs. per sq. ft. live load.
Deflection—Limited to span in inches divided by 240 for live load only.

ALLOWABLE CLEAR SPANS FOR VARIOUS SIZES AND SPECIES OF JOISTS *(cont.)*

Floor Joists

Span (feet and inches)

Species	Grade	2 x 8		2 x 10		2 x 12	
		16" oc	24" oc	16" oc	24" oc	16" oc	24" oc
Douglas Fir-Larch	2 & better	13-1	11-3	16-9	14-5	20-4	17-6
	3	10-7	8-8	13-6	11-0	16-5	13-5
Douglas Fir South	2 & better	12-0	10-6	15-3	13-4	18-7	16-3
	3	10-3	8-4	13-1	10-8	15-11	13-0
Hem-Fir	2 & better	12-3	10-0	15-8	12-10	19-1	15-7
	3	9-5	7-8	12-0	9-10	14-7	11-11
Mountain Hemlock	2 & better	11-4	9-11	14-6	12-8	17-7	15-4
	3	9-7	7-10	12-3	10-0	14-11	12-2
Mountain Hemlock-Hem-Fir	2 & better	11-4	9-11	14-6	12-8	17-7	15-4
	3	9-5	7-8	12-0	9-10	14-7	11-11
Western Hemlock	2 & better	12-3	10-6	15-8	13-4	19-1	16-3
	3	9-11	8-1	12-8	10-4	15-5	12-7

Species	Grade						
Engelmann Spruce Alpine Fir (Engelmann Spruce-Lodgepole Pine)	2 & better 3	11-2 8-6	9-1 6-11	14-3 10-10	11-7 8-10	17-3 13-2	14-2 10-9
Lodgepole Pine	2 & better 3	11-8 9-1	9-7 7-5	14-11 11-7	12-3 9-5	18-1 14-1	14-11 11-6
Ponderosa Pine-Sugar Pine (Ponderosa Pine-Lodgepole Pine)	2 & better 3	11-4 8-8	9-3 7-1	14-5 11-1	11-9 9-1	17-7 13-6	14-4 11-0
White Woods (Western Woods)	2 & better 3	11-0 8-6	9-0 6-11	14-0 10-10	11-6 8-10	17-0 13-2	14-0 10-9
Idaho White Pine	2 & better 3	11-0 8-6	9-0 6-11	14-0 10-10	11-6 8-10	17-1 13-2	14-0 10-9
Western Cedars	2 & better 3	11-0 9-1	9-7 7-5	14-0 11-6	12-3 9-5	17-0 14-0	14-11 11-6

Design Criteria:
Strength—10 lbs. per sq. ft. dead load plus 40 lbs. per sq. ft. live load.
Deflection—Limited to span in inches divided by 360 for live load only.

LENGTHS OF COMMON RAFTERS

Feet of Run	2 in 12 Inclination (set saw at) 9° 28' 12.17 in. per ft. of run	2½ in 12 Inclination (set saw at) 11° 46' 12.26 in. per ft. of run	3 in 12 Inclination (set saw at) 14° 2' 12.37 in. per ft. of run	3½ in 12 Inclination (set saw at) 16° 16' 12.5 in. per ft. of run	4 in 12 Inclination (set saw at) 18° 26' 12.65 in. per ft. of run	4½ in 12 Inclination (set saw at) 20° 33' 12.82 in. per ft. of run
4	4' 11/16"	4' 1 1/32"	4' 1 15/32"	4' 2"	4' 2 19/32"	4' 3 9/32"
5	5' 27/32"	5' 1 5/16"	5' 1 27/32"	5' 2 1/2"	5' 3 1/4"	5' 4 3/32"
6	6' 1 1/32"	6' 1 9/16"	6' 2 7/32"	6' 3"	6' 3 29/32"	6' 4 15/16"
7	7' 1 3/16"	7' 1 13/16"	7' 2 19/32"	7' 3 1/2"	7' 4 9/16"	7' 5 3/4"
8	8' 1 3/8"	8' 2 3/32"	8' 2 31/32"	8' 4"	8' 5 3/32"	8' 6 9/16"
9	9' 1 17/32"	9' 2 7/16"	9' 3 11/32"	9' 4 1/2"	9' 5 27/32"	9' 7 3/8"
10	10' 1 23/32"	10' 2 19/32"	10' 3 23/32"	10' 5"	10' 6 1/2"	10' 8 7/32"
11	11' 1 7/8"	11' 2 7/8"	11' 4 1/16"	11' 5 1/2"	11' 7 5/32"	11' 9 1/32"
12	12' 2 1/32"	12' 3 1/8"	12' 4 7/16"	12' 6"	12' 7 13/16"	12' 9 27/32"
13	13' 2 7/32"	13' 3 3/8"	13' 4 13/16"	13' 6 1/2"	13' 8 15/32"	13' 10 21/32"
14	14' 2 3/8"	14' 3 21/32"	14' 5 3/16"	14' 7"	14' 9 3/32"	14' 11 1/2"
15	15' 2 9/16"	15' 3 29/32"	15' 5 9/16"	15' 7 1/2"	15' 9 3/4"	16' 5/16"
16	16' 2 23/32"	16' 4 5/32"	16' 5 15/16"	16' 8"	16' 10 13/32"	17' 1 1/8"

These lengths are to be added to those shown above when run involves inches.

Inches of Run						
1/4"	1/4"	1/4"	1/4"	1/4"	1/4"	9/32"
1/2"	1/2"	1/2"	1/2"	17/32"	17/32"	7/32"
1"	1"	1"	1 1/32"	1 1/32"	1 1/16"	1 1/16"
2"	2 1/32"	2 1/32"	2 1/16"	2 3/32"	2 3/32"	2 1/8"
3"	3 1/16"	3 1/16"	3 3/32"	3 1/8"	3 5/32"	3 7/32"
4"	4 1/16"	4 3/32"	4 1/8"	4 5/32"	4 7/32"	4 9/32"
5"	5 1/16"	5 1/8"	5 5/32"	5 7/32"	5 9/32"	5 11/32"
6"	6 3/32"	6 1/8"	6 3/16"	6 1/4"	6 5/16"	6 13/32"
7"	7 3/32"	7 5/32"	7 7/32"	7 9/32"	7 3/8"	7 15/32"
8"	8 1/8"	8 3/16"	8 1/4"	8 11/32"	8 7/16"	8 17/32"
9"	9 1/8"	9 3/16"	9 1/4"	9 3/8"	9 15/32"	9 5/8"
10"	10 5/32"	10 7/32"	10 5/16"	10 7/16"	10 17/32"	10 11/16"
11"	11 5/32"	11 1/4"	11 11/32"	11 15/32"	11 19/32"	11 3/4"

4-31

LENGTHS OF COMMON RAFTERS (cont.)

Feet of Run	5 in 12 Inclination (set saw at) 22° 37' 13.00 in. per ft. of run	5½ in 12 Inclination (set saw at) 24° 37' 13.20 in. per ft. of run	6 in 12 Inclination (set saw at) 26° 34' 13.42 in. per ft. of run	6½ in 12 Inclination (set saw at) 28° 27' 13.65 in. per ft. of run	7 in 12 Inclination (set saw at) 30° 15' 13.89 in. per ft. of run	7½ in 12 Inclination (set saw at) 32° 0' 14.15 in. per ft. of run	8 in 12 Inclination (set saw at) 33° 41' 14.42 in. per ft. of run
4	4' 4"	4' 4 13/16"	4' 5 11/16"	4' 6 19/32"	4' 7 9/16"	4' 8 19/32"	4' 9 11/16"
5	5' 5"	5' 6"	5' 7 1/8"	5' 8 1/4"	5' 9 15/32"	5' 10 3/4"	6' 1/8"
6	6' 6"	6' 7 3/32"	6' 8 1/2"	6' 9 29/32"	6' 11 11/32"	7' 29/32"	7' 2 17/32"
7	7' 7"	7' 8 13/32"	7' 9 15/16"	7' 11 9/16"	8' 1 7/32"	8' 3 1/16"	8' 4 15/16"
8	8' 8"	8' 9 19/32"	8' 11 3/8"	9' 1 7/32"	9' 3 1/8"	9' 5 7/32"	9' 7 3/8"
9	9' 9"	9' 10 13/16"	10' 25/32"	10' 2 27/32"	10' 5"	10' 7 11/32"	10' 9 25/32"
10	10' 10"	11' 0"	11' 2 7/32"	11' 4 1/2"	11' 6 29/32"	11' 9 1/2"	12' 7/32"
11	11' 11"	12' 1 3/32"	12' 3 5/8"	12' 6 5/32"	12' 8 13/16"	12' 11 21/32"	13' 2 5/8"
12	13' 0"	13' 2 13/32"	13' 5 1/32"	13' 7 13/16"	13' 10 11/16"	14' 1 13/16"	14' 5 1/32"
13	14' 1"	14' 3 19/32"	14' 6 15/32"	14' 9 15/32"	15' 9/16"	15' 3 31/32"	15' 7 15/32"
14	15' 2"	15' 4 13/16"	15' 7 7/8"	15' 11 1/8"	16' 2 15/32"	16' 6 1/8"	16' 9 7/8"
15	16' 3"	16' 6"	16' 9 5/16"	17' 3/4"	17' 4 11/32"	17' 8 1/4"	18' 5/16"
16	17' 4"	17' 7 7/32"	17' 10 23/32"	18' 2 13/32"	18' 6 1/4"	18' 10 13/32"	19' 2 23/32"

These lengths are to be added to those shown above when run involves inches.

Inches of Run							
1/4"	5/16"	5/16"	9/32"	9/32"	9/32"	9/32"	9/32"
1/2"	5/8"	19/32"	9/16"	9/16"	9/16"	9/16"	17/32"
1"	1 7/32"	1 3/16"	1 5/32"	1 1/8"	1 1/8"	1 3/32"	1 3/32"
2"	2 13/32"	2 11/32"	2 5/16"	2 9/32"	2 1/4"	2 7/32"	2 5/32"
3"	3 19/32"	3 17/32"	3 15/32"	3 13/32"	3 3/8"	3 5/16"	3 1/4"
4"	4 13/16"	4 23/32"	4 5/8"	4 9/16"	4 15/32"	4 13/32"	4 11/32"
5"	6"	5 29/32"	5 25/32"	5 11/16"	5 19/32"	5 1/2"	5 13/32"
6"	7 7/32"	7 1/32"	6 15/16"	6 13/16"	6 23/32"	6 19/32"	6 1/2"
7"	8 13/32"	8 1/4"	8 1/8"	7 31/32"	7 13/16"	7 23/32"	7 9/16"
8"	9 5/8"	9 7/16"	9 1/4"	9 1/8"	8 15/16"	8 13/16"	8 21/32"
9"	10 13/16"	10 19/32"	10 13/32"	10 1/4"	10 5/32"	9 29/32"	9 3/4"
10"	1'	11 25/32"	11 9/16"	11 3/8"	11 3/16"	11"	10 27/32"
11"	1' 1 7/32"	1' 31/32"	1' 23/32"	1' 1/2"	1' 5/16"	12 3/32"	11 29/32"

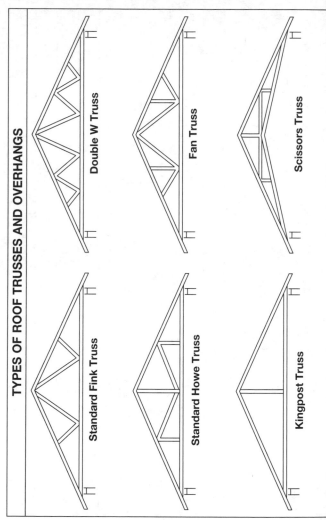

TYPES OF ROOF TRUSSES AND OVERHANGS

Double W Truss

Fan Truss

Scissors Truss

Standard Fink Truss

Standard Howe Truss

Kingpost Truss

4-34

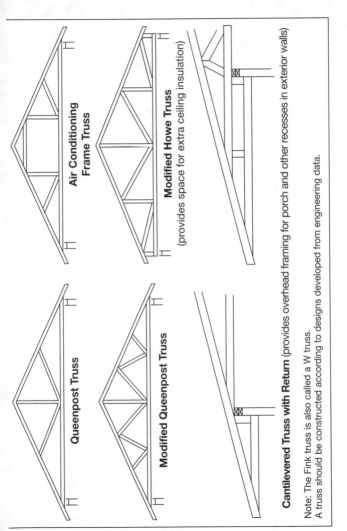

Queenpost Truss

Air Conditioning Frame Truss

Modified Queenpost Truss

Modified Howe Truss
(provides space for extra ceiling insulation)

Cantilevered Truss with Return (provides overhead framing for porch and other recesses in exterior walls)

Note: The Fink truss is also called a W truss.
A truss should be constructed according to designs developed from engineering data.

4-35

RAFTER LAYOUT TERMS

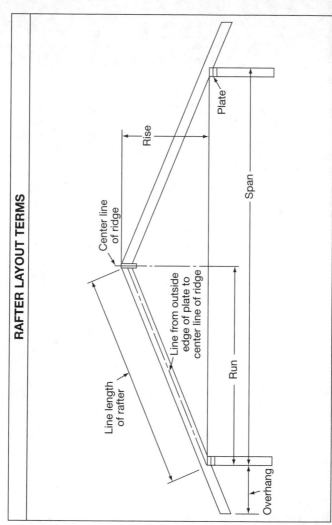

Center line of ridge

Rise

Plate

Span

Line length of rafter

Line from outside edge of plate to center line of ridge

Run

Overhang

4-36

ALLOWABLE UNIFORM LOADS FOR W STEEL BEAMS

Designation (wt./ft.)	Nominal Size (dp. x wd.)	Span in Feet									
		8'	10'	12'	14'	16'	18'	20'	22'	24'	26'
W8 x 10	8 x 4	15.6	12.5	10.4	8.9	7.8	6.9	—	—	—	—
W8 x 13	8 x 4	19.9	15.9	13.3	11.4	9.9	8.8	—	—	—	—
W8 x 15	8 x 4	23.6	18.9	15.8	13.5	11.8	10.5	—	—	—	—
W8 x 18	8 x 5$\frac{1}{4}$	30.4	24.3	20.3	17.4	15.2	13.5	—	—	—	—
W8 x 21	8 x 5$\frac{1}{4}$	36.4	29.1	24.3	20.8	18.2	16.2	—	—	—	—
W8 x 24	8 x 6$\frac{1}{2}$	41.8	33.4	27.8	23.9	20.9	18.6	—	—	—	—
W8 x 28	8 x 6$\frac{1}{2}$	48.6	38.9	32.4	27.8	24.3	21.6	—	—	—	—
W10 x 22	10 x 5$\frac{3}{4}$	—	—	30.9	26.5	23.2	20.6	18.6	16.9	—	—
W10 x 26	10 x 5$\frac{3}{4}$	—	—	37.2	31.9	27.9	24.8	22.3	20.3	—	—
W10 x 30	10 x 5$\frac{3}{4}$	—	—	43.2	37.0	32.4	28.8	25.9	23.6	—	—
W12 x 26	12 x 6$\frac{1}{2}$	—	—	—	—	33.4	29.7	26.7	24.3	22.3	20.5
W12 x 30	12 x 6$\frac{1}{2}$	—	—	—	—	38.6	34.3	30.9	28.1	25.8	23.8
W12 x 35	12 x 6$\frac{1}{2}$	—	—	—	—	45.6	40.6	36.5	33.2	30.4	28.1

Note: Loads are given in kips (1 kip = 1000 lb.)

LOAD LIMITS FOR WOOD GIRDERS

Wood Girders	Safe Load in lb. for Spans From 6 to 10 Feet				
Size	6 Ft.	7 Ft.	8 Ft.	9 Ft.	10 Ft.
6 x 8 Solid	8,306	7,118	6,220	5,539	4,583
6 x 8 Built-up	7,359	6,306	5,511	4,908	4,062
6 x 10 Solid	11,357	10,804	9,980	8,887	7,997
6 x 10 Built-up	10,068	9,576	8,844	7,878	7,086
8 x 8 Solid	11,326	9,706	8,482	7,553	6,250
8 x 8 Built-up	9,812	8,408	7,348	6,544	5,416
8 x 10 Solid	15,487	14,782	13,608	12,116	10,902
8 x 10 Built-up	13,424	12,768	11,792	10,504	9,448

SOUND INSULATION OF DOUBLE WALLS

Wall Detail	Description	STC Rating
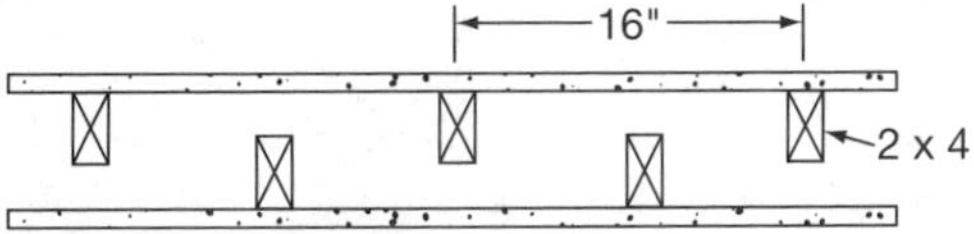16" 2 x 4	$^1/_2$-inch gypsum wallboard	45
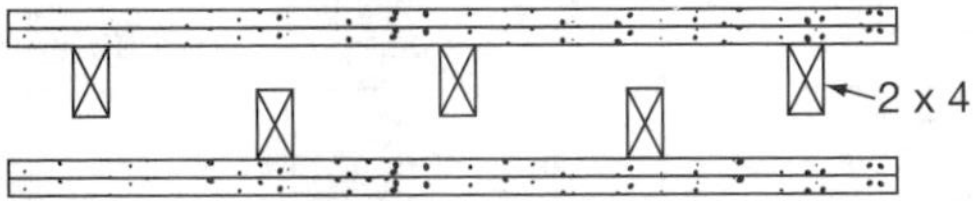2 x 4	$^5/_8$-inch gypsum wallboard (double layer each side)	45
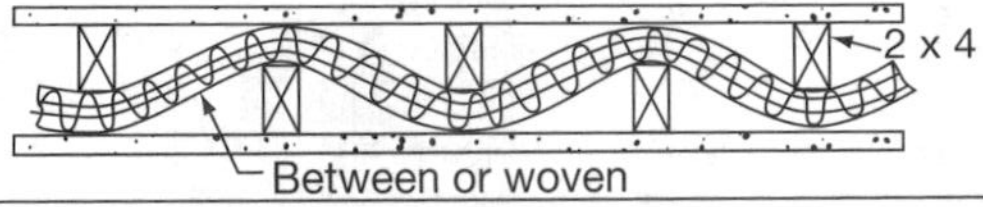2 x 4 Between or woven	$^1/_2$-inch gypsum wallboard $1^1/_2$-inch fibrous insulation	49
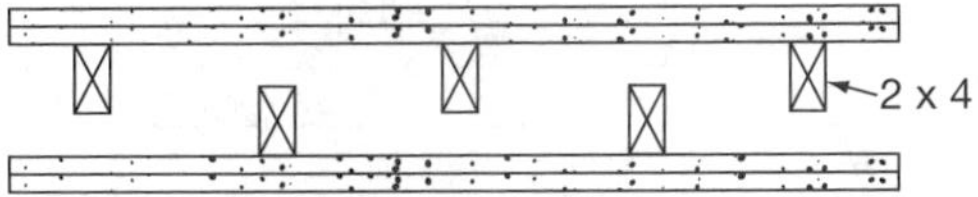2 x 4	$^1/_2$-inch sound deadening board (nailed) $^1/_2$-inch gypsum wallboard (laminated)	50

FIRE RETARDATION AND SOUND TRANSMISSION RATING

Partitions – Wood Framing (load-bearing)

	Fire Rating	Ref.	Description
Single Layer	45-min.	U.L. Design No.1 – 45 min.	$1/2$" fire-shield gypsum wallboard, nailed both sides 2" x 4" studs, 16" o.c.
Single Layer	1-hour	U.L. Design No. 5 – 1 hr.	$5/8$" fire-shield gypsum wallboard or fire-shield M-R board nailed both sides 2" x 4" wood studs, 16" o.c.
Single Layer	1-hour	U.L. Design No. 25 – 1 hr.	$5/8$" fire-shield gypsum wallboard nailed both sides 2" x 4" wood studs, 24" o.c.
Single Layer	1-hour	F.M. Design WP-90 – 1 hr.	$5/8$" fire-shield monolithic Durasan, vertically applied to 2" x 4" studs spaced 24" o.c. secured at joints with 6d nails spaced 7" o.c. and at intermediate studs with $3/8$" x $3/8$" bead of MC adhesive.
Single Layer (resilient)	1-hour	Based on O.S.U. T-3376 & U.L. Design No. 5 – 1 hr.	$5/8$" fire-shield gypsum wallboard, screw applied to resilient furring channel, spaced 24" o.c. one side only, on 2" x 4" studs spaced 16" o.c. Other side $5/8$" fire-shield gypsum wallboard nailed direct to studs.
Single Layer (resilient)	1-hour	Based on O.S.U. T-3376	$5/8$" fire-shield gypsum wallboard, screw applied to resilient furring channel, spaced 24" o.c. one side only, on 2" x 4" studs spaced 16" o.c. Other side $5/8$" fire-shield gypsum wallboard screw attached at 16" spacing, 3" fiberglass in stud cavity.
Single Layer (resilient)	1-hour	O.S.U. T-3376	$5/8$" fire-shield gypsum wallboard, screw applied to resilient furring channels 24" o.c. nailed to both sides of 2" x 4" studs spaced 16" o.c.

Category	Rating	Design	Description
Double Layer	1-hour	F.M. Design WP-147 – 1 hr.	$\frac{1}{2}$" fire-shield wallboard or Durasan laminated to $\frac{1}{4}$" gypsum wallboard nailed to both sides 2" x 4" studs spaced 16" o.c.
Double Layer	2-hour	Based on U.L. Design No. 4 – 2 hr.	$\frac{5}{8}$" fire-shield gypsum wallboard base layer nail applied to 2" x 4" wood stud spaced 16" o.c. Face layer $\frac{5}{8}$" Fire-shield gypsum wallboard laminated and nail applied
Exterior Walls	1-hour	F.M. Design WP-78 – 1 hr. U.L. Design	$\frac{5}{8}$" fire-shield gypsum wallboard nailed horizontally to inside face of 2" x 4" wood studs 16" o.c.; $\frac{1}{2}$" gypsum sheathing nailed to outside face of studs. Siding $\frac{3}{8}$" woodrock.
Exterior Walls	2-hour	No. 23 – 2 hr.	Two layers $\frac{5}{8}$" fire-shield gypsum wallboard nailed horiz. or vert. to inside face of 2" x 4" wood studs 16" o.c.; $\frac{1}{2}$" gypsum sheathing nailed to outside face of studs, brick veneer facing.
Double Layer with Deciban Sound Deadening Board	Non-rated		$\frac{1}{2}$" Deciban nail applied both sides 2" x 4" wood studs 16" o.c. Face layer $\frac{1}{2}$" gypsum wallboard laminated.
Double Layer with Deciban Sound Deadening Board	Non-rated		$\frac{1}{2}$" Deciban nail applied to both sides 2" x 4" wood studs. Face 16" o.c. fire-topped face layer. $\frac{5}{8}$" gypsum wallboard laminated.
Double Layer with Deciban Sound Deadening Board	1-hour	U.L. Design No. 17 – 1 hr.	$\frac{1}{2}$" Deciban nail applied to both sides 2" x 3" wood studs, staggered 16" o.c. on 2" x 3" plates spaced 1" apart. Face layer $\frac{5}{8}$" fire-Shield gypsum wallboard nail applied.
Double Layer with Deciban Sound Deadening Board	1-hour	U.L. Design No. 26 – 1 hr.	$\frac{1}{2}$" Deciban nail applied both sides 2" x 3" wood studs, staggered 24" o.c. on 2" x 3" plates spaced 1" apart. Face layer $\frac{1}{2}$" fire-shield gypsum wallboard nail applied.

4-41

FIRE RETARDATION AND SOUND TRANSMISSION RATING *(cont.)*

Partitions – Steel Framing

		Fire Rating	Ref.	Description
Unbalanced 2½" Studs		1-hour	F.M. Design WP-66	½" fire-shield (Monolithic Durasan) vertically applied to 2½" screw stud. Double layer one side, single layer on the other. Base layer screw attached, face layer and single layer screwed at edges, adhesively attached along center.
		1-hour	Based on F.M. Design WP-66	½" fire-shield gypsum wallboard screw attached vertically to both sides, 2½" screw studs spaced 24" o.c. Second layer screw attached vertically to one side only.
		1-hour	Based on F.M. Design WP-66	½" fire-shield gypsum wallboard screw attached vertically to both sides, 2½" screw studs spaced 24" o.c. Second layer screw attached vertically to one side only and 3" fiberglass in cavity.
		1½-hour	Based on O.S.U. T-3240	⅝" fire-shield gypsum wallboard screw attached vertically to both sides, 2½" screw studs spaced 24" o.c. Second layer screw attached vertically to one side only.
		1½-hour	Based on O.S.U. T-3240	⅝" fire-shield gypsum wallboard screw attached vertically to both sides, 2½" screw studs spaced 24" o.c. Second layer screw attached vertically to one side only and 3" fiberglass in cavity.
3⅝" Studs		1-hour	Based on F.M. Design WP-66	½" fire-shield gypsum wallboard screw attached vertically to both sides, 3⅝" screw studs spaced 24" o.c. Second layer screw attached vertically to one side only.
		1-hour	Based on F.M. Design WP-66	½" fire-shield gypsum wallboard screw attached vertically to both sides, 3⅝" screw studs spaced 24" o.c. Second layer screw attached vertically to one side only and 3" fiberglass in cavity.
		1½-hour	O.S.U. T-3240	⅝" fire-shield gypsum wallboard screw attached vertically to both sides, 3⅝" screw studs spaced 24" o.c. Second layer laminated vertically and screwed to one side only.

Stud Group	Fire Rating	Reference	Description
1⁵/₈" Studs Single Layer	1-hour	O.S.U. T-3296	$5/8$" fire-shield gypsum wallboard screw attached vertically to both sides, $1^5/_8$" screw studs 24" o.c.
	1-hour	Based on O.S.U. T-3296	$5/8$" fire-shield gypsum wallboard screw attached vertically to both sides, $1^5/_8$" screw studs 24" o.c. with 1" fiberglass in cavity.
2¹/₂" Studs Single Layer	1-hour	Based on O.S.U. T-3296	$5/8$" fire-shield gypsum wallboard screw attached vertically to both sides, $2^1/_2$" screw studs 24" o.c.
	1-hour	Based on O.S.U. T-3296	$5/8$" fire-shield gypsum wallboard screw attached vertically to both sides, $2^1/_2$" screw studs 24" o.c. with 3" fiberglass in cavity.
	1-hour	F.M. Design WP-51	$1/2$" fire-shield gypsum wallboard screw attached vertically to both sides, $2^1/_2$" screw studs 24" o.c., 2" mineral wool in stud cavity.
3⁵/₈" Studs Single Layer	1-hour	F.M. Design WP-45	$5/8$" fire-shield gypsum wallboard screw attached horizontally to both sides, $3^5/_8$ screw studs 24" o.c. Wallboard joints staggered.
	1-hour	Based on O.S.U. T-1770	$5/8$" fire-shield gypsum wallboard screw attached vertically to both sides, $3^5/_8$" screw studs 24" o.c. with 3" fiberglass in cavity.
	45-min.	Based on F.M. Design WP-51	$1/2$" fire-shield gypsum wallboard screw attached vertically to both sides, $3^5/_8$" screw studs 24" o.c., 2" fiberglass in cavity.
	1-hour	Based on F.M. Design WP-51	$1/2$" fire-shield gypsum wallboard screw attached vertically to both sides, $3^5/_8$" screw studs 24" o.c., 2" mineral wool in stud cavity.
	1-hour	O.S.U. T-1770	$5/8$" fire-shield gypsum wallboard screw attached vertically to both sides, $3^5/_8$" screw studs 24" o.c.

4-43

FIRE RETARDATION AND SOUND TRANSMISSION RATING (cont.)

Partitions – Steel Framing (cont.)

	Fire Rating	Ref.	Description
Unbalanced 3⅝" Studs	1½-hour	Based on O.S.U. T-3240	⅝" fire-shield gypsum wallboard screw attached vertically to both sides, 3⅝" screw studs 24" o.c. Second layer laminated vertically and screwed to one side only and 2" fiberglass or min. wool in cavity.
Double Layer 2½" Studs	1-hour	F.M. Design WP-152	½" fire-shield wallboard or Durasan laminated to ¼" wallboard screw attached both sides, 2½" screw studs spaced 24" o.c.
	1-hour	Based on F.M. Design WP-152	½" fire-shield wallboard or Durasan laminated to ¼" wallboard screw attached both sides, 2½" screw studs spaced 24" o.c. with 2" fiberglass in cavity.
	2-hour	O.S.U. T-3370	Two layers ½" fire-shield gypsum wallboard screw attached vertically both sides, 2½" screw studs spaced 24" o.c. Vertical joints staggered.
	2-hour	F.M. Design WP-47	First layer ½" fire-shield gypsum wallboard screw attached vertically both sides, 2½" screw studs spaced 24" o.c. Second layer screw attached horizontally both sides.
	2-hour	Based on O.S.U. T-3370	Two layers ½" fire-shield gypsum wallboard screw attached vertically both sides, 2½" screw studs spaced 24" o.c. Vertical joints staggered and 3" fiberglass in cavity.
	2-hour	Based on F.M. Design WP-47 O.S.U. T-1771	First layer ⅝" fire-shield gypsum wallboard screw attached vertically both sides, 2½" screw studs spaced 24" o.c. Second layer screw attached horizontally both sides and 3" fiberglass in cavity.

2-hour	Based on O.S.U. T-3370	Two layers $1/2$" fire-shield gypsum wallboard screw attached vertically both sides, $3^5/8$" screw studs spaced 24" o.c. Vertical joints staggered.
2-hour	Based on O.S.U. T-3370	Two layers $1/2$" fire-shield gypsum wallboard screw attached vertically both sides, $3^5/8$" screw studs spaced 24" o.c. Vertical joints staggered and 3" fiberglass in cavity.
2-hour	O.S.U. T-1771	First layer $5/8$" fire-shield gypsum wallboard screw attached vertically both sides, $3^5/8$" screw studs spaced 24" o.c. Second layer laminated vertically both sides.
2-hour	Based on F.M. Design WP-47	First layer $5/8$" fire-shield gypsum wallboard screw attached vertically both sides, $3^5/8$" screw studs spaced 24" o.c. Second layer screw attached horizontally both sides and 3" fiberglass in cavity.

Double Layer $3^5/8$" Studs

DOOR AND FRAME SELECTOR (Type of Door Frame)

Door Weight	Gauge Jamb Studs	Bracing Over Header	Knock Down Alum.	Knock Down Steel	Fixed Steel	Requires Mechanical Closure	Height* Stud Size			
							1⁵/₈"	2¹/₂"	3⁵/₈"	4"
	25		x				NA**	10'	14'	15'
	25			x			NA	10'	14'	15'
	25				x		NA	10'	14'	15'
Up	25	x	x				10'	12'	16'	17'
to	25	x		x			10'	12'	16'	17'
50 lbs.	25	x			x		10'	12'	16'	17'
	20	x	x				—	16'	21'	22'
	20	x		x			—	16'	21'	22
	20	x			x		—	16'	21'	22'

4-46

Weight	Stud									
50 lbs. to 80 lbs.	25	x				x	12'	10'	14'	15'
	25	x				x		10'	14'	15'
	25	x		x				10'	14'	15'
	25	x	x				10'	12'	14'	15'
	20		x			x		10'	14'	15'
	20	x		x		x		10'	14'	15'
	20	x				x		10'	14'	15'
	20	x		x				12'	16'	17'
	20	x			x			12'	16'	17'
	20 dbl.	x			x			16'	21'	22'
	20 dbl.			x				16'	21'	22'
80 lbs. to 120 lbs.	25	x			x	x		12'	16'	17'
	20	x	x					12'	16'	17'
	20				x	x		12'	16'	17'
	20 dbl.	x		x	x			16'	21'	22'

*See partition height table for stud spacing.
**NA-Not allowed.

4-47

DETAILS OF METAL WALL CONSTRUCTION

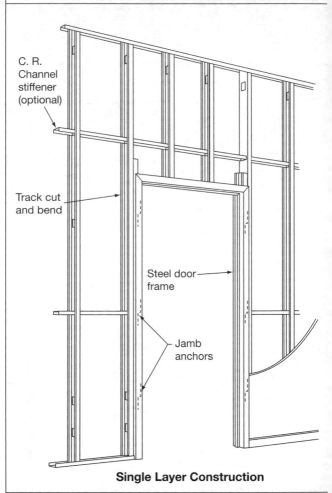

C. R. Channel stiffener (optional)

Track cut and bend

Steel door frame

Jamb anchors

Single Layer Construction

USING METAL STUDS AND TRACKS

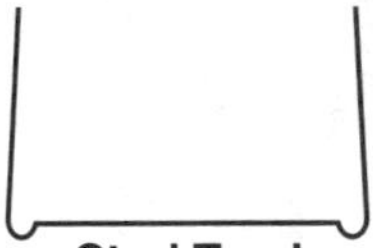

24" O.C. (nominal) continuously

Stud Track
(cross-section)

Screw Stud
(cross-section)

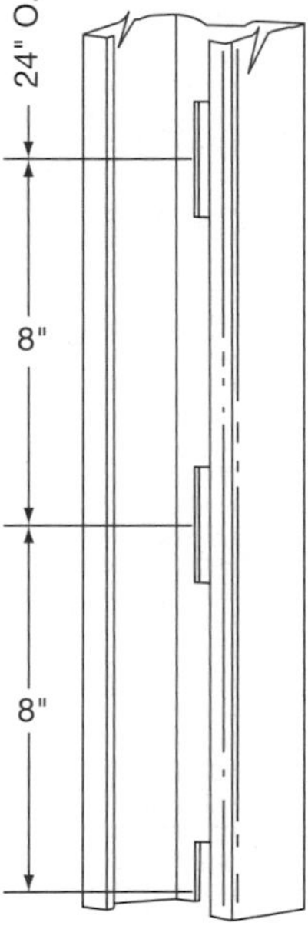

8"

8"

Resilient
furring
channel

Wallboard

Screw
stud

Track

Screw Stud

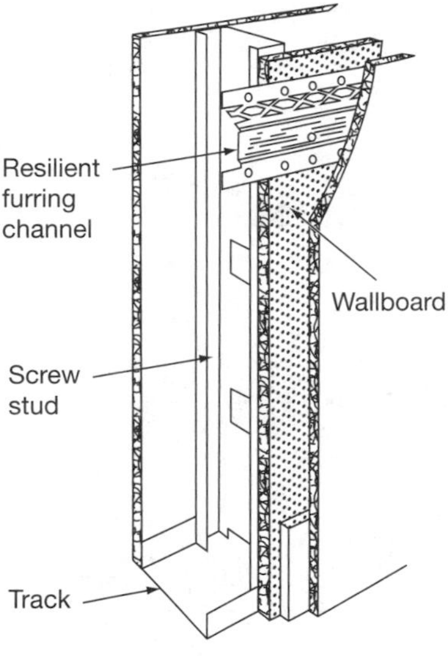

**Resilient Furring Channel
with Gypsum Wallboard**

4-49

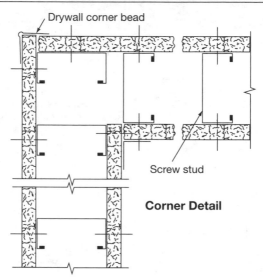

Drywall corner bead

Screw stud

Corner Detail

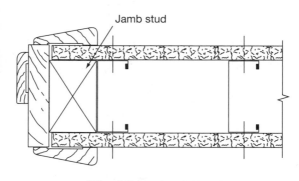

Jamb stud

Wood Door Jamb Detail

JAMBS — METAL WALL CONSTRUCTION

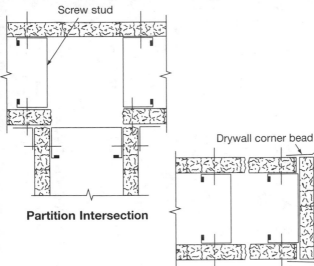

Screw stud

Partition Intersection

Drywall corner bead

Partition End Detail

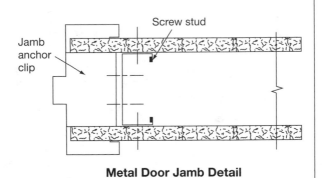

Jamb anchor clip

Screw stud

Metal Door Jamb Detail

PLYWOOD MARKING

APA

Panel grade ——— RATED SHEATHING

Span rating ——— 32/16 1/2 INCH ——— Thickness

SIZED FOR SPACING

Exposure durability ——— EXPOSURE 1
classification

——— 000 ——— Mill number

National Research ——— NRB-108
Board report number

Typical grade-trademark which is stamped on all plywood panels manufactured in compliance with national plywood standard.

PLYWOOD GRADES AND USAGE

Protected or Interior Use

Grade Designation	Description & Common Uses	Typical Trademarks
APA Rated Sheathing EXP 1 or 2	Specially designed for subflooring and wall and roof sheathing, but can also be used for a broad range of other construction and industrial applications. Can be manufactured as conventional veneered plywood, as a composite, or as a nonveneered panel. For special engineered applications, including high load requirements and certain industrial uses, veneered panels conforming to PS 1 may be required. Specify Exposure 1 when long construction delays are anticipated. Common thicknesses: $5/16$, $3/8$, $7/16$, $1/2$, $5/8$, $3/4$.	APA RATED SHEATHING 32/16 $1/2$ INCH SIZED FOR SPACING EXPOSURE 1 000 NRB-108
APA Structural I & II Rated Sheathing EXP 1	Unsanded all-veneer PS 1 plywood grades for use where strength properties are of maximum importance: structural diaphragms, box beams, gusset plates, stressed-skin panels, containers, pallet bins. Made only with exterior glue (Exposure 1). STRUCTURAL I more commonly available. Common thicknesses: $5/16$, $3/8$, $1/2$, $5/8$, $3/4$.	APA RATED SHEATHING STRUCTURAL I 24/0 $3/8$ INCH SIZED FOR SPACING EXPOSURE 1 PS 1-74 C-D INT/EXT GLUE 000 NRB-108

4-53

PLYWOOD GRADES AND USAGE (cont.)

Protected or Interior Use

Grade Designation	Description & Common Uses	Typical Trademarks
APA Rated Sturd-I-Floor EXP 1 or 2	For combination subfloor-underlayment. Provides smooth surface for application of resilient floor covering and possesses high concentrated and impact load resistance. Can be manufactured as conventional veneered plywood, as a composite, or as a non-veneered panel. Available square edge or tongue-and-groove. Specify Exposure 1 when long construction delays are anticipated. Common thicknesses: $5/8$ ($19/32$), $3/4$ ($23/32$).	**APA** RATED STURD-I-FLOOR **24/oc** 23/32 INCH SIZED FOR SPACING T&G NET WIDTH 47-1/2 EXPOSURE 1 000 NRB-108
APA Rated Sturd-I-Floor 48 oc (2-4-1) EXP 1	For combination subfloor-underlayment on 32- and 48-inch spans and for heavy timber roof construction. Provides smooth surface for application of resilient floor coverings and possesses high concentrated and impact load resistance. Manufactured only as conventional veneered plywood and only with exterior glue (Exposure 1). Available square edge or tongue-and-groove. Thickness: $1\frac{1}{8}$.	**APA** RATED STURD-I-FLOOR **48 oc** 1-1/8 INCH (2-4-1) SIZED FOR SPACING EXPOSURE 1 T&G 000 INT EXT GLUE NRB-108 FHA-UM-66

Exterior Use

APA Rated Sheathing EXT	Exterior sheathing panel for subflooring and wall and roof sheathing, siding on service and farm buildings, crating, pallets, pallet bins, cable reels, etc. Manufactured as conventional veneered plywood. Common thicknesses: $5/16$, $3/8$, $1/2$, $5/8$, $3/4$.	**APA** RATED SHEATHING **48/24** 3/4 INCH SIZED FOR SPACING EXTERIOR _____ 000 _____ NRB-108
APA Structural I & II Rated Sheathing EXT	For engineered applications in construction and industry where resistance to permanent exposure to weather or moisture is required. Manufactured only as conventional veneered PS I plywood. Unsanded. STRUCTURAL I more commonly available. Common thicknesses: $5/16$, $3/8$, $1/2$, $5/8$, $3/4$.	**APA** RATED SHEATHING STRUCTURAL I **42/20** 5/8 INCH SIZED FOR SPACING EXTERIOR _____ 000 _____ PS 1-74 C-C NRB-108
APA Rated Sturd-I-Floor EXT	For combination subfloor-underlayment under resilient floor coverings where severe moisture conditions may be present, as in balcony decks. Possesses high concentrated and impact load resistance. Manufactured only as conventional veneered plywood. Available square edge or tongue-and-groove. Common thicknesses: $5/8$ ($19/32$), $3/4$ ($23/32$).	**APA** RATED STURD-I-FLOOR **20 OC** 19/32 INCH SIZED FOR SPACING EXTERIOR _____ 000 _____ NRB-108

Note: Specific grades, thicknesses, constructions and exposure durability classifications may be in limited supply in some areas. Check with your supplier before specifying. Specify Performance-Rated Panels by thickness and Span Rating.

SOFTWOOD PLYWOOD VENEER GRADES

N	Smooth surface "natural finish" veneer. Select, all heartwood or all sapwood. Free of open defects. Allows not more than 6 repairs, wood only, per 4 x 8 panel, made parallel to grain and well matched for grain and color.
A	Smooth, paintable. Not more than 18 neatly made repairs, boat, sled, or router type, and parallel to grain, permitted. May be used for natural finish in less demanding applications.
B	Solid surface. Shims, circular repair plugs, and tight knots to 1 inch across grain permitted. Some minor splits permitted.
C Plugged	Improved C veneer with splits limited to $\frac{1}{8}$ inch width and knotholes and borer holes limited to $\frac{1}{4} \times \frac{1}{2}$ inch. Admits some broken grain. Synthetic repairs permitted.
C	Tight knots to $1\frac{1}{2}$ inch. Knotholes to 1 inch across grain and some to $1\frac{1}{2}$ inch if total width of knots and knotholes is within specified limits. Synthetic or wood repairs. Discoloration and sanding defects that do not impair strength permitted. Limited splits allowed. Stitching permitted.
D	Knots and knotholes to $2\frac{1}{2}$ inch width across grain and $\frac{1}{2}$ inch larger within specified limits. Limited splits allowed. Stitching permitted. Limited to Interior, Exposure 1 and Exposure 2 panels.

EXPOSED PLYWOOD PANEL SIDING

Minimum Thickness[1]	Minimum No. of Plies	Stud Spacing (inches) Plywood Siding Applied Direct to Studs or Over Sheathing
$3/8$"	3	16[1]
$1/2$"	4	24

Thickness of grooved panels is measured at bottom of grooves.
[1]May be 24 inches if plywood siding applied with face grain perpendicular to studs or over one of the following:
a) 1-inch board sheathing.
b) $15/32$-inch plywood sheathing or
c) $3/8$-inch plywood sheathing with face grain of sheathing perpendicular to studs.

ALLOWABLE SPANS FOR EXPOSED PARTICLEBOARD PANEL SIDING

Grade	Stud Spacing (inches)	Minimum Thickness (inches)		Exterior Ceilings and Soffits
		Siding		
		Direct to Studs	Continuous Support	Direct to Supports
2-M-W	16	$3/8$	$5/16$	$5/16$
	24	$1/2$	$5/16$	$3/8$
2-M-1	16	$5/8$	$3/8$	—
2-M-2				
2-M-3	24	$3/4$	$3/8$	—

ALLOWABLE SPANS FOR PARTICLEBOARD SUBFLOOR AND COMBINED SUBFLOOR-UNDERLAYMENT

Grade	Thickness (inches)	Maximum Spacing of Supports	
		Subfloor	Combined Subfloor-Underlayment
2-M-W	$1/2$	16	—
	$5/8$	20	16
	$3/4$	24	24
2-M-3	$3/4$	20	20

Note: All panels are continuous over two or more spans. Uniform deflection limitation: $1/360$ of the span under 100 psf minimum load. Edges shall have tongue-and-groove joints or shall be supported with blocking. The tongue-and-groove panels are installed with the long dimension perpendicular to supports.

ALLOWABLE SPANS FOR PLYWOOD COMBINATION SUBFLOOR-UNDERLAYMENT

Plywood Continuous Over Two or More Spans and Face Grain Perpendicular to Supports

Identification	Spacing of Joists (inches)			
	16	20	24	48
Species Group[1]	Thickness in inches			
1	$1/2$	$5/8$	$3/4$	—
2, 3	$5/8$	$3/4$	$7/8$	—
4	$3/4$	$7/8$	1	—
Span Rating[2]	16 o.c.	20 o.c.	24 o.c.	48 o.c.

Spans limited to value shown because of possible effect of concentrated loads. Allowable uniform load based on deflection of $1/360$ of span is 125 pounds per square foot (psf), except allowable total uniform load for $1\frac{1}{8}$-inch plywood over joists spaced 48 inches on center is 65 psf. Plywood edges shall have approved tongue-and-groove joints or shall be supported with blocking, unless $1/4$-inch minimum thickness underlayment is installed, or finish floor is $3/4$-inch wood strip. If wood strips are perpendicular to supports, thicknesses shown for 16-inch and 20-inch spans may be used on 24-inch span.

[1]Applicable to all grades of sanded exterior-type plywood.
[2]Applicable to underlayment grade and C-C (plugged).

CLASSIFICATION OF SOFTWOOD PLYWOODS RATES SPECIES FOR STRENGTH AND STIFFNESS

Group 1

Apitong	Douglas Fir 1	Maple, Sugar	Pine, South Loblolly
Beech, American	Kapur	Pine Caribbean	Longleaf Shortleaf
Birch Sweet Yellow	Keruing Larch, Western	Ocote	Slash Tanoak

Group 2

Cedar, Port Orford	Fir *(cont.)* Pacific Silver White	Lauan *(cont.)* Tangile White Lauan	Pine *(cont.)* Virginia Western White
Cypress	Hemlock, Western	Maple, Black	Spruce Red Sitka
Douglas Fir 2	Lauan Almon Bagtikan Mayapis Red Lauan	Mengkulang Meranti, Red Mersawa Pine Pond Red	Sweetgum Tamarack Yellow-poplar
Fir California Red Grand Noble			

Group 3

Adler, Red	Fir, Subalpine	Pine Jack Lodgepole Ponderosa Spruce	Redwood
Birch, Paper	Hemlock, Eastern		Spruce Black Engelmann White
Cedar, Alaska	Maple, Bigleaf		

Group 4

Aspen Bigtooth Quaking	Cedar Incense Western Red	Cottonwood *(cont.)* Black (Western Poplar)	Pine Eastern White Sugar
Cativo	Cottonwood Eastern		

Group 5

Basswood	Fir, Balsam	Poplar, Balsam	

Note: Group 1 represents strongest woods.

OPTIMUM SPACING OF SPRING CLIPS FOR CEILING

25 Gauge Studs — Maximum Heights

Stud Spacing	1 5/8" Stud		2 1/2" Stud		3 1/4" Stud		3 5/8" Stud		4" Stud	
	1/2" GWB	5/8" GWB	1/2" GWB	5/8" GWB	1/2" GWB	5/8" GWB	1/2" GWB	5/8" GWB	1/2" GWB	5/8" GWB
12" o.c.	12' 0"	12' 4"	16' 2"	16' 6"	19' 6"	19' 10"	21' 0"	21' 4"	22' 0"	22' 4"
16" o.c.	11' 0"	11' 7"	14' 8"	15' 5"	17' 10"	18' 4"	19' 5"	19' 11"	20' 8"	20' 10"
24" o.c.	10' 0"	10' 10"	13' 5"	14' 3"	15' 10"	16' 7"	17' 3"	18' 2"	18' 5"	19' 2"

20 Gauge Studs — Maximum Heights

Stud Spacing	2 1/2" Stud		3 1/4" Stud		3 5/8" Stud		4" Stud	
	1/2" WB	5/8" WB	1/2" WB	5/8" WB	1/2" WB	5/8" WB	1/2" WB	5/8" WB
12" o.c.	18' 9"	19' 0"	22' 8"	22' 11"	24' 8"	24' 10"	26' 7"	26' 10"
16" o.c.	17' 9"	18' 0"	21' 7"	21' 9"	23' 5"	23' 7"	25' 3"	25' 6"
24" o.c.	15' 9"	16' 0"	18' 11"	19' 2"	20' 5"	20' 8"	22' 0"	22' 3"

CHAPTER 5
Roofing

TYPICAL ASPHALT ROLLS

Product	Approximate Shipping Weight		Sqs. per Pkg.	Length	Width	Side or End Lap	Top Lap	Exposure	UL Listing
	Per Roll	Per Square							
Mineral surface roll	75# to 90#	75# to 90#	1	36' to 38'	36"	6"	2" to 4"	32" to 34"	C
Mineral surface roll (double coverage)	55# to 70#	110# to 140#	½	36'	36"	6"	19"	17"	C
Smooth surface roll	40# to 65#	40# to 65#	1	36'	36"	6"	2"	34"	None
Saturated felt (non-perforated)	60#	15# to 30#	2 to 4	72' to 144'	36"	4" to 6"	2" to 19"	17" to 34"	None

TYPICAL ASPHALT SHINGLES

Product	Configur-ation	Per Square			Size			Exposure	UL Listing
		Approx. Shipping Weight	Shingles	Bundles	Width	Length			
Self-sealing random-tab strip shingle Multi-thickness	Various edge, surface texture and application treatments	285# to 390#	66 to 90	4 or 5	11 1/2" to 14"	36" to 40"		4" to 6"	A or C Many wind resistant
Self-sealing random-tab strip shingle Single-thickness	Various edge, surface texture and application treatments	250# to 300#	66 to 80	3 or 4	12" to 13 1/4"	36" to 40"		5" to 5 5/8"	A or C Many wind resistant

Product	Configuration	Approximate Weight per Square	Shingles per Square	Bundles per Square	Width	Length	Exposure	UL Rating
Self-sealing square-tab strip shingle — Three-tab	Two-tab or Four-tab	215# to 325#	66 to 80	3 or 4	12" to 13 1/4"	36" to 40"	5" to 5 5/8"	A or C All wind resistant
	Three-tab	215# to 300#	66 to 80	3 or 4	12" to 13 1/4"	36" to 40"	5" to 5 5/8"	A or C All wind resistant
Self-sealing square-tab strip shingle — No-cutout	Various edge and surface texture treatments	215# to 290#	66 to 81	3 or 4	12" to 13 1/4"	36" to 40"	5" to 5 5/8"	A or C All wind resistant
Individual interlocking shingle — Basic design	Several design variations	180# to 250#	72 to 120	3 or 4	18" to 22 1/4"	20" to 22 1/2"	—	C Many wind resistant

HEATING TEMPERATURES OF ASPHALT

ASTM D-312 Type No.	Asphalt Type	Maximum Heating Temperature (°F)
I	Dead Level	475
II	Flat	500
III	Steep	525
IV	Special steep	525

NAILER SPACING

Incline (in.)	Smooth	Gravel	Cap Sheet
0-1	Not required	Not required	Not required
1-2	Not required	20' face to face	20' face to face
2-3	20' face to face	10' face to face	10' face to face
3-4	10' face to face	Not recommended	4' face to face
4-6	4' face to face	Not recommended	4' face to face

NAILER SPACING FOR VARIOUS ROOF TYPES

Asphalt/Cap Sheet Roofs

Incline (in.)	Nailer Spacing (D)	Asphalt Type
0 – $\frac{1}{2}$	Not required	II
$\frac{1}{2}$ – 1	Not required	III
1 – 2	20' face to face	III
2 – 3	10' face to face	III
3 – 6	4' face to face	IV

Asphalt/Smooth-Surfaced Roofs

Incline (in.)	Nailer Spacing (D)	Asphalt Type
0 – $\frac{1}{2}$	Not required	II
$\frac{1}{2}$ – 1	Not required	II
1 – 2	Not required	III
2 – 3	20' face to face	III
3 – 4	10' face to face	IV
4 – 6	4' face to face	IV

Asphalt/Gravel-Surfaced Roofs

Incline (in.)	Nailer Spacing (D)	Asphalt Type
0 – $\frac{1}{2}$	Not required	II
$\frac{1}{2}$ – 1	Not required	III
1 – 2	20' face to face	III
2 – 3	10' face to face	III

RECOMMENDED NAIL SIZES

Purpose	Nail Length (in.)
Roll roofing materials on new decks	1
Strip or individual shingles on new decks	$1\frac{1}{4}$
Reroofing over old asphalt roofing materials	$1\frac{1}{4} - 1\frac{1}{2}$
Reroofing over old wooden shingles	$1\frac{3}{4}$
Reroofing over 300-pound or heavier asphalt shingles	$1\frac{1}{2} - 1\frac{3}{4}$

DESIGN OF GUTTER SYSTEMS

Areas for Pitched Roofs

Pitch	Factor*
Level to 3 in./ft.	1.00
4 to 5 in./ft.	1.05
6 to 8 in./ft.	1.10
9 to 11 in./ft.	1.20
12 in./ft.	1.30

*To determine the design area, multiply the plan area by this factor.

DIMENSIONS OF STANDARD LEADERS

Type	Area (sq. in.)	Leader Sizes (in.)
Plain round	7.07	3
	12.57	4
	19.63	5
	28.27	6
Corrugated round	5.94	3
	11.04	4
	17.72	5
	25.97	6
Polygon octagonal	6.36	3
	11.30	4
	17.65	5
	25.40	6
Square corrugated	3.80	$1^3/_4 \times 2^1/_4$ (2)
	7.73	$2^3/_8 \times 3^1/_4$ (3)
	11.70	$2^3/_4 \times 4^1/_4$ (4)
	18.75	$3^3/_4 \times 5$ (5)
Plain rectangular	3.94	$1^3/_4 \times 2^1/_4$
	6.00	2×3
	8.00	2×4
	12.00	3×4
	20.00	$3^3/_4 \times 4^3/_4$
	24.00	4×6
SPS pipe	7.38	3
	12.72	4
	20.00	5
	28.88	6
Cast-iron pipe	7.07	3
	12.57	4
	19.64	5
	28.27	6

SHINGLE COVERAGE

Weather Exposure

Length x Thickness	3½"	4"	5"	5½"	6"	6½"	7"	7½"
	Approximate Coverage (in sq. ft.) of 1 Square (4 bundles) of Shingles							
16" x 5/2	70	80	90	100	—	—	—	—
18" x 5/2¼	—	72½	81½	90½	100*	—	—	—
24" x 4/2	—	—	—	—	73½	80	86½	93	100*

*Maximum exposure recommended for roofs.

SHINGLE EXPOSURE

	No. 1 Blue Label			No. 2 Red Label			No. 3 Black Label		
Pitch	Length (in.)								
	16	18	24	16	18	24	16	18	24
3/12 – 4/12	3¾	4¼	5¾	3½	4	5½	3	3½	5
4/12 and steeper	5	5½	7½	4	4½	6½	3½	4	5½

ATTACHMENT OF TILES TO SHEATHING

Field Tile Nailing

Roof Slope	Solid Sheathing with Battens	Solid Sheathing Without Battens[1]	Nailing for Perimeter, Tile and Tile on Cantilevered Areas[2]
3/12 to and including 5/12	Not required	Every tile	Every tile
Above 5/12 to less than 12/12	Every tile every other row	Every tile	Every tile
12/12 and over	Every tile	Every tile	Every tile

[1]Battens are required for slopes exceeding 7/12.
[2]Perimeter nailing areas including three tile courses but not less than 36 inches from either side of hips or ridges and edges of eaves and gable rakes. In special wind areas, as designated by the building official, additional fastenings might be required.

LATH SPACING AND SLATE LENGTH

Spacing of Lath (O.C.) (in.)	Length of Slate (in.)
$10\frac{1}{2}$	24
$9\frac{1}{2}$	22
$8\frac{1}{2}$	20
$7\frac{1}{2}$	18

TAPER SIZES

Length of Sheets (in.)	Increase in Width (in.)	Length of Sheets (in.)	Increase in Width (in.)
24	$1/4$	60	$5/8$
30	$5/16$	66	$11/16$
36	$3/8$	72	$3/4$
42	$7/16$	84	$7/8$
48	$1/2$	96	1
54	$9/16$		

EXPOSURE FOR SLOPING ROOFS

Length of Slate (in.)	Exposure (in.)*
24	$10 1/2$
22	$9 1/2$
20	$8 1/2$
18	$7 1/2$
16	$6 1/2$
14	$5 1/2$
12	$4 1/2$
10	$3 1/2$

*Slope 8 to 20 inches per foot, 3-inch lap.

SHAKE COVERAGE					
	Weather exposure				
	5"	5½"	7½"	8½"	10"
Length x Thickness, Shake Type	Approximate Coverage (in sq. ft.) of 1 Square of Shakes[a]				
18" x ½", hand-split and resawn mediums[b]	—	55[c]	75[d]	—	—
18" x ¾", hand-split and resawn heavies[b]	—	55[c]	75[d]	—	—
18" x ⅝", tapersawn	—	55[c]	75[d]	—	—
24" x ⅜", hand-split	50[e]	—	75[c]	—	—
24" x ½", hand-split and resawn mediums	—	—	75[c]	85	100[d]
24" x ¾", hand-split and resawn heavies	—	—	75[c]	85	100[d]
24" x ⅝", tapersawn	—	—	75[c]	85	100[d]
24" x ½", tapersawn	—	—	75[c]	85	100[d]
18" x ⅜", straightsplit	—	65[c]	90[d]	—	—
24" x ⅜", straightsplit	—	—	75[c]	85	100[d]

15" starter-finish course: Use as supplemental with shakes applied not to exceed 10" of weather exposure.

[a]All coverage based on an average ½" spacing between shakes.

[b]5 bundles cover 100 sq. ft. roof area when used as starter-finish course at 10" weather exposure; 7 bundles cover 100 sq. ft. roof area at 7½" weather exposure. See [a].

[c]Maximum recommended weather exposure for 3-ply roof construction.

[d]Maximum recommended weather exposure for 2-ply roof construction.

[e]Maximum recommended weather exposure.

SHINGLE FASTENERS

Shingle or Shake Type	Nail Type	Min. Length (in.)
Shingles—new roof		
16" and 18" shingles	3d box	$1\frac{1}{3}$
24" shingles	4d box	$1\frac{1}{2}$
Shakes—new roof		
18" straightsplit	5d box	$1\frac{3}{4}$
18" and 24" hand-split and resawn	6d box	2
24" tapersplit	5d box	$1\frac{3}{4}$
18" and 24" tapersawn	6d box	2

AVERAGE WEIGHT OF SLATE PER SQUARE

Slate Thickness (in.)	Sloping Roof With 3" Lap (lb. per square)	Flat Roof Without Lap (lb. per square)
$\frac{3}{16}$	700	240
$\frac{3}{16}$	750	250
$\frac{1}{4}$	1000	335
$\frac{3}{8}$	1500	500
$\frac{1}{2}$	2000	675
$\frac{3}{4}$	3000	1000
1	4000	1330
$1\frac{1}{4}$	5000	1670
$1\frac{1}{2}$	6000	2000
$1\frac{3}{4}$	7000	
2	8000	

SCHEDULE FOR STANDARD ³⁄₁₆" THICK SLATE

Size of Slate (in.)	Slates per Square	Exposure with 3" Lap (in.)	Nails per Square (lb. oz.)	
26 x 14	89	11½	1	0
24 x 16	86	10½	1	0
24 x 14	98	10½	1	2
24 x 13	106	10½	1	3
24 x 12	114	10½	1	5
24 x 11	125	10½	1	7
22 x 14	108	9½	1	4
22 x 13	117	9½	1	5
22 x 12	126	9½	1	7
22 x 11	138	9½	1	9
22 x 10	152	9½	1	12
20 x 14	121	8½	1	6
20 x 13	132	8½	1	8
20 x 12	141	8½	1	10
20 x 11	154	8½	1	12
20 x 10	170	8½	1	15
20 x 9	189	8½	2	3
18 x 14	137	7½	1	9
18 x 13	148	7½	1	11
18 x 12	160	7½	1	13
18 x 11	175	7½	2	0
18 x 10	192	7½	2	3
18 x 9	213	7½	2	7

SCHEDULE FOR STANDARD ³⁄₁₆" THICK SLATE *(cont.)*

Size of Slate (in.)	Slates per Square	Exposure with 3" Lap (in.)	Nails per Square (lb. oz.)	
16 x 14	160	6½	1	13
16 x 12	184	6½	2	2
16 x 11	201	6½	2	5
16 x 10	222	6½	2	8
16 x 9	246	6½	2	13
16 x 8	277	6½	3	2
14 x 12	218	5½	2	8
14 x 11	238	5½	2	11
14 x 10	261	5½	3	3
14 x 9	291	5½	3	5
14 x 8	327	5½	3	12
14 x 7	374	5½	4	4
12 x 10	320	4½	3	10
12 x 9	355	4½	4	1
12 x 8	400	4½	4	9
12 x 7	457	4½	5	3
12 x 6	533	4½	6	1
11 x 8	450	4	5	2
11 x 7	515	4	5	14
10 x 8	515	3½	5	14
10 x 7	588	3½	7	4
10 x 6	686	3½	7	13

CONSTRUCTION TECHNIQUES FOR ROOF VENTILATION

No. 1 red cedar shingles or shakes

Louvered vent at each end of attic

Roof rafter

Continuous screened vent

Insulation

Air flow

Air flow

5-15

FREE AREA VENTILATION GUIDE

Length (in feet)	Square inches of ventilation required for attic areas. Width (in feet)											
	20	22	24	26	28	30	32	34	36	38	40	42
20	192	211	230	250	269	288	307	326	346	365	384	403
22	211	232	253	275	296	317	338	359	380	401	422	444
24	230	253	276	300	323	346	369	392	415	438	461	484
26	250	275	300	324	349	374	399	424	449	474	499	524
28	269	296	323	349	376	403	430	457	484	511	538	564
30	288	317	346	374	403	432	461	490	518	547	576	605
32	307	338	369	399	430	461	492	522	553	584	614	645
34	326	359	392	424	457	490	522	555	588	620	653	685
36	346	380	415	449	484	518	553	588	622	657	691	726
38	365	401	438	474	511	547	584	620	657	693	730	766
40	384	422	461	499	538	576	614	653	691	730	768	806
42	403	444	484	524	564	605	645	685	726	766	806	847
44	422	465	507	549	591	634	676	718	760	803	845	887
46	442	486	530	574	618	662	707	751	795	839	883	927
48	461	507	553	599	645	691	737	783	829	876	922	968
50	480	528	576	624	672	720	768	816	864	912	960	1008

PLYWOOD ROOF DECKING

Identi-fication Index	Plywood Thickness (inches)	Maximum Span (inches)	Unsupported Edge-Max. Length (inches)	Allowable Live Loads, psf									
				Spacing of Supports Center to Center (inches)									
				12	16	20	24	30	32	36	42	48	60
12/0	5/16	12	12	150									
16/0	5/16, 3/8	16	16	160	75								
20/0	5/16, 3/8	20	20	190	105	65							
24/0	3/8, 1/2	24	24	250	140	95	50						
32/16	1/2, 5/8	32	28	385	215	150	95	50	40				
42/20	5/8, 3/4, 7/8	42	32		330	230	145	90	75	50	35		
48/24	3/4, 7/8	48	36			300	190	120	105	65	45	35	
2•4•1	1 1/8	72	48				390	245	215	135	100	75	45
1 1/8 Grp. 1 and 2	1 1/8	72	48				305	195	170	105	75	55	35
1 1/4 Grp. 3 and 4	1 1/4	72	48				355	225	195	125	90	85	40

HIP-AND-VALLEY CONVERSIONS

Rise, Inches per Foot of Horizontal Run	4	5	6	7	8	9	10	11	12	14	16	18
Pitch, Degrees Pitch, Fractions	18°26' $\frac{1}{6}$	22°37' $\frac{5}{24}$	26°34' $\frac{1}{4}$	30°16' $\frac{7}{24}$	33°41' $\frac{1}{3}$	36°52' $\frac{3}{8}$	39°48' $\frac{5}{12}$	42°31' $\frac{11}{24}$	45° $\frac{1}{2}$	49°24' $\frac{7}{12}$	53°8' $\frac{2}{3}$	56°19' $\frac{3}{4}$
Conversion Factor	1.452	1.474	1.500	1.524	1.564	1.600	1.642	1.684	1.732	1.814	1.944	2.062
Horizontal Length in Feet												
1	1.5	1.5	1.5	1.5	1.6	1.6	1.6	1.7	1.7	1.8	1.9	2.1
2	2.9	2.9	3.0	3.0	3.1	3.2	3.3	3.4	3.5	3.6	3.9	4.1
3	4.4	4.4	4.5	4.6	4.7	4.8	4.9	5.1	5.2	5.4	5.8	6.2
4	5.8	5.9	6.0	6.1	6.3	6.4	6.6	6.7	6.9	7.3	7.8	8.2
5	7.3	7.4	7.5	7.6	7.8	8.0	8.2	8.4	8.7	9.1	9.7	10.3
6	8.7	8.8	9.0	9.1	9.4	9.6	9.9	10.1	10.4	10.9	11.7	12.4
7	10.2	10.3	10.5	10.7	10.9	11.2	11.5	11.8	12.1	12.7	13.6	14.4

8	11.6	11.8	12.0	12.2	12.5	12.8	13.1	13.5	13.9	14.5	15.6	16.5
9	13.1	13.3	13.5	13.7	14.1	14.4	14.8	15.2	15.6	16.3	17.5	18.6
10	14.5	14.7	15.0	15.2	15.6	16.0	16.4	16.8	17.3	18.1	19.4	20.6
20	29.0	29.5	30.0	30.5	31.3	32.0	32.8	33.7	34.6	36.3	38.9	41.2
30	43.6	44.2	45.0	45.7	46.9	48.0	49.3	50.5	52.0	54.4	58.3	61.9
40	58.1	59.0	60.0	61.0	62.6	64.0	65.7	67.4	69.3	72.6	77.8	82.5
50	72.6	73.7	75.0	76.2	78.2	80.0	82.1	84.2	86.6	90.7	97.2	103.1
60	87.1	88.4	90.0	91.4	93.8	96.0	98.5	101.0	103.9	108.8	116.6	123.7
70	101.6	103.2	105.0	106.7	109.5	112.0	114.9	117.9	121.2	127.0	136.1	144.3
80	116.2	117.9	120.9	121.9	125.1	128.0	131.4	134.7	138.6	145.1	155.5	165.0
90	130.7	132.7	135.0	137.2	140.8	144.0	147.8	151.6	155.9	163.3	175.0	185.6
100	145.2	147.4	150.0	152.4	156.4	160.0	164.2	168.4	173.2	181.4	194.4	206.2

CONVERSION TABLE — HORIZONTAL TO SLOPE AREAS

Rise, Inches per Foot of Horizontal Run	1	2	3	4	5	6	7	8	9	10	11	12
Pitch, Fractions	$1/24$	$1/12$	$1/8$	$1/6$	$5/24$	$1/4$	$7/24$	$1/3$	$3/8$	$5/12$	$11/24$	$1/2$
Conversion Factor	1.004	1.014	1.031	1.054	1.083	1.118	1.157	1.202	1.250	1.302	1.356	1.414
Horizontal Length in Feet												
1	1.0	1.0	1.0	1.1	1.1	1.1	1.2	1.2	1.3	1.3	1.4	1.4
2	2.0	2.0	2.1	2.1	2.2	2.2	3.2	2.4	2.5	2.6	2.7	2.8
3	3.0	3.0	3.1	3.2	3.2	3.2	3.5	3.6	3.8	3.9	4.1	4.2
4	4.0	4.1	4.1	4.2	4.3	4.5	4.6	4.8	5.0	5.2	5.4	5.7
5	5.0	5.1	5.2	5.3	5.4	5.6	5.8	6.0	6.3	6.5	6.8	7.1
6	6.0	6.1	6.2	6.3	6.5	6.7	6.9	7.2	7.5	7.8	8.1	8.5
7	7.0	7.1	7.2	7.4	7.6	7.8	8.1	8.4	8.8	9.1	9.5	9.9
8	8.0	8.1	8.3	8.4	8.7	8.9	9.3	9.6	10.0	10.4	10.8	11.3
9	9.0	9.1	9.3	9.5	9.7	10.1	10.4	10.8	11.3	11.7	12.2	12.7
10	10.0	10.1	10.3	10.5	10.8	11.2	11.6	12.0	12.5	13.0	13.6	14.1
20	20.1	20.3	20.6	21.1	21.7	22.4	23.1	24.0	25.0	26.0	27.1	28.3

30	30.1	30.4	31.0	31.6	32.5	33.5	34.7	36.1	37.5	39.1	40.7	42.4
40	40.2	40.6	41.2	42.2	43.3	44.7	46.3	48.1	50.0	52.1	54.2	56.6
50	50.2	50.7	51.6	52.7	54.2	55.9	57.8	60.1	62.5	65.1	67.8	70.7
60	60.2	60.8	61.9	63.2	65.0	67.1	69.4	72.1	75.0	78.1	81.4	84.8
70	70.3	71.0	72.2	73.8	75.8	78.3	81.0	84.1	87.5	91.1	94.9	99.0
80	80.3	81.1	82.5	84.3	86.6	89.4	92.6	96.2	100.0	104.2	108.5	113.1
90	90.4	91.3	92.8	94.9	97.5	100.6	104.1	108.2	112.5	117.2	122.0	127.3
100	100.4	101.4	103.1	105.4	108.3	111.8	115.7	120.2	125.0	130.2	135.6	141.4
200	200.8	202.8	206.2	210.8	216.6	223.6	231.4	240.4	250.0	260.4	271.2	282.8
300	301.2	304.2	309.3	316.2	324.9	335.4	347.1	360.6	375.0	390.6	406.8	424.2
400	401.6	405.6	412.4	421.6	433.2	447.2	462.8	480.8	500.0	520.8	542.4	565.6
500	502.0	507.0	515.5	527.0	541.5	559.0	578.5	601.0	625.0	651.0	678.0	707.0
600	602.4	608.4	618.6	632.4	649.8	670.8	694.2	721.2	750.0	781.2	813.6	848.4
700	702.8	709.8	721.7	737.8	758.1	782.6	809.9	841.4	875.0	911.4	949.2	989.8
800	803.2	811.2	824.8	843.2	864.4	894.4	925.6	961.6	1000.0	1041.6	1084.8	1131.2
900	903.6	912.6	927.9	948.6	974.7	1006.2	1041.3	1081.8	1125.0	1171.8	1220.4	1272.6
1000	1004.0	1014.0	1031.0	1054.0	1083.0	1118.0	1157.0	1202.0	1250.0	1302.0	1356.0	1414.0

ROOF DEAD LOADS

Material		Weight, psf
Aluminum (including laps):	Flat	Corrugated (1½ and 2½ in.)
12 American or B&S gauge	1.2	—
14	0.9	1.1
16	0.7	0.9
18	0.6	0.7
20	0.5	0.6
22	—	0.4
Galvanized Steel (including laps):		Corrugated (2½ and 3 in.)
12 U.S. standard gauge	4.5	4.9
14	3.3	3.6
16	2.7	2.9
18	2.2	2.4
20	1.7	1.8
22	1.4	1.5
24	1.2	1.3
26	0.9	1.0
Other Types of Decking (per inch of thickness):		
Concrete plank	6.5	
Insulrock	2.7	
Petrical	2.7	
Porex	2.7	
Poured gypsum	6.5	
Tectum	2.0	
Vermiculite concrete	2.6	
Corrugated Asbestos (¼ in.)	3.0	
Felt:		
3-ply	1.5	
3-ply with gravel	5.5	
5-ply	2.5	
5-ply with gravel	6.5	
Insulation (per inch of thickness):		
Expanded polystyrene	0.2	
Fiberglass, rigid	1.5	
Loose	0.5	
Roll Roofing	1.0	

CHAPTER 6
Drywall and Plaster

FRAME SPACING FOR DRYWALL CONSTRUCTION		
Single-ply Gypsum Board Thickness (in.)	Application to Framing	Maximum o.c. Spacing of Framing (in.)
Ceilings:		
$3/8$	Perpendicular*	16
$1/2$	Perpendicular	16
$1/2$	Parallel*	16
$5/8$	Parallel	16
$1/2$	Perpendicular*	24
$5/8$	Perpendicular	24
Sidewalls:		
$3/8$	Perpendicular or parallel	16
$1/2$	Perpendicular* or parallel	24
$5/8$	Perpendicular or parallel	24

*On ceilings to receive a water-base texture material, either hand or spray applied, install gypsum board perpendicular to framing and increase board thickness from $3/8$ to $1/2$ in. for 16 in. o.c. framing and from $1/2$ to $5/8$ in. for 24 in. o.c. framing. $3/8$ in. should not support thermal insulation.

MAXIMUM FASTENER SPACING – DRYWALL

Framing	Type Const.	Type Fastener	Location	Max. Fastener Spacing			
				Gypsum Panels		Gypsum Base	
				in.	mm	in.	mm
Wood	single layer	nails	ceilings	7	178	7	178
			sidewalls	8	203	8	203
		screws	ceilings	12	305	12	305
			sidewalls	16	406	12	305
		screws with RC-1 channels	ceilings	12	305	12	305
			sidewalls	12	305	12	305
	base layer of double layer — both layers mechanically attached	nails	ceilings	24	610	24	610
			sidewalls	24	610	24	610
		screws	ceilings	24	610	24	610
			sidewalls	24	610	24	610
	face layer of double layer — both layers mechanically attached	nails	ceilings	7	178	7	178
			sidewalls	8	203	8	203
		screws	ceilings	12	305	12	305
			sidewalls	16	406	12	305
	base layer of double layer — face layer adhesively attached	nails	ceilings	7	178	7	178
			sidewalls	8	203	8	203
		screws	ceilings	12	305	12	305
			sidewalls	16	406	12	305
	face layer of double layer — face layer adhesively attached	nails	ceilings	16" o.c. at ends and edges — 1 field fastener per frame member at mid-width of board	406 mm o.c. at ends and edges and 1 field fastener per frame member	same as for gypsum panels	same as for gypsum panels
			sidewalls	fasten top and bottom as required	fasten top and bottom as required	same as for gypsum panels	same as for gypsum panels

Steel							
	single layer	screws	ceilings	12	305	12	305
			sidewalls	16	406	12	305
	base layer of double layer — both layers mechanically attached	screws	ceilings	16	406	16	406
			sidewalls	24	610	24	610
	face layer of double layer — both layers mechanically attached	screws	ceilings	12	305	12	305
			sidewalls	16	406	12	305
	base layer of double layer — face layer adhesively attached	screws	ceilings	12	305	12	305
			sidewalls	16	406	12	305
	face layer of double layer — face layer adhesively attached	screws	ceilings	16" o.c. at ends and edges — 1 field fastener per frame member at mid-width of board	406 mm o.c. at ends and edges plus 1 field fastener per frame member	same as for gypsum panels	same as for gypsum panels
			sidewalls	fasten top and bottom as required	fasten top and bottom as required	same as for gypsum panels	same as for gypsum panels

DRYING TIME FOR JOINT COMPOUND UNDER TAPE								
RH	RH = Relative Humidity D = Days (24 hr.) H = Hours							
98%	53 D	38 D	26 D	18 D	12 D	9 D	6 D	4½ D
97%	37 D	26 D	18 D	12 D	9 D	6 D	4½ D	3¼ D
96%	28 D	21 D	14 D	10 D	7 D	5 D	3½ D	2½ D
95%	25 D	17 D	12 D	8 D	6 D	4 D	2¾ D	2 D
94%	20 D	14 D	10 D	7 D	5 D	3¼ D	2¼ D	41 H
93%	18 D	12½ D	9 D	6 D	4 D	2¾ D	2 D	36 H
92%	15 D	11 D	8 D	5 D	3½ D	2½ D	44 H	32 H
91%	14 D	10 D	7 D	4¾ D	3¼ D	2¼ D	40 H	29 H
90%	13 D	9 D	6 D	4½ D	3 D	49 H	36 H	26 H
85%	10 D	6 D	4 D	3 D	2 D	34 H	25 H	18 H
80%	7 D	4¾ D	3¼ D	2¼ D	38 H	27 H	19 H	14 H
70%	4½ D	3½ D	2¼ D	38 H	26 H	19 H	14 H	10 H
60%	3½ D	2½ D	42 H	29 H	20 H	14 H	10 H	8 H
50%	3 D	2 D	36 H	24 H	17 H	12 H	9 H	6 H
40%	2½ D	44 H	29 H	20 H	14 H	10 H	7 H	5 H
30%	2¼ D	38 H	26 H	18 H	12 H	9 H	6 H	4½ H
20%	2 D	34 H	23 H	16 H	11 H	8 H	5½ H	4 H
10%	42 H	30 H	21 H	14 H	10 H	7 H	5 H	3½ H
0	38 H	28 H	19 H	13 H	9 H	6 H	4½ H	3 H
°F	32°	40°	50°	60°	70°	80°	90°	100°
°C	0°	4°	10°	16°	21°	27°	32°	38°

ALLOWABLE CARRYING LOADS FOR ANCHOR BOLTS

Type Fastener		Size	Allowable Load	
			1/2" Wallboard	5/8" Wallboard
Hollow wall screw anchors		1/8" dia. short	50 lbs.	—
		3/16" dia. short	65 lbs.	—
		1/4", 5/16", 3/8" dia. short	65 lbs.	—
		3/16" dia. long	—	90 lbs.
		1/4", 5/16", 3/8" dia. long	—	95 lbs.
Common toggle bolts		1/8" dia.	30 lbs.	90 lbs.
		3/16" dia.	60 lbs.	120 lbs.
		1/4", 5/16", 3/8" dia.	80 lbs.	120 lbs.

6-5

MIXING PLASTER					
	Cement	Lime	Sand	Yield	Mix by Volume
Round Sand (Dune or River Sand)					
Exterior	50 kg	25 kg	220 kg	165 L	1:1:5
Interior	50 kg	50 kg	345 kg	250 L	1:2:7.5
Sharp Sand (Quarried Sand)					
Exterior	50 kg	—	320 kg	210 L	1:7
Interior	50 kg	25 kg	400 kg	270 L	1:1:9

CHAPTER 7
Flooring and Tile

HARDWOOD FLOORING GRADES

Flooring is bundled by averaging the lengths. A bundle may include pieces from 6 inches under to 6 inches over the nominal length of the bundle. No piece is shorter than 9 inches. Quantity with length under 4 feet held to stated percentage of total footage in any one shipment of item. ¾ inch added to face length when measuring length of each piece.

Unfinished Oak Flooring

CLEAR (Plain or Quarter Sawn)**
Best appearance, best grade, most uniform color, limited small character marks.
Bundles 1¼ ft. and up. Average length 3¾ ft.

SELECT AND BETTER (Special Order)
A combination of Clear and Select grades.

SELECT (Plain or Quarter Sawn)**
Excellent appearance, limited character marks, unlimited sound sap. Bundles 1¼ ft. and up. Average length 3¼ ft.

NO. 1 COMMON
Variegated appearance, light and dark colors, knots, flags, worm holes and other character marks allowed to provide a variegated appearance, after imperfections are filled and finished. Bundles 1¼ ft. and up. Average length 2¾ ft.

NO. 2 COMMON (Red & White may be mixed)
Rustic appearance, all wood characteristics of species, a serviceable economical floor after knot holes, worm holes, checks and other imperfections are filled and finished. Bundles 1¼ ft. and up. Average length 2¼ ft.

Beech, Birch, Hard Maple

FIRST GRADE WHITE HARD MAPLE (Special Order)
Same as First Grade except face all bright sapwood.

FIRST GRADE RED BEECH & BIRCH (Special Order)
Same as First Grade except face all red heartwood.

FIRST GRADE
Best appearance, natural color variation, limited
character marks, unlimited sap.
Bundles 2 ft. and up; 2 ft. & 3 ft. bundles
up to 33% footage.

SECOND AND BETTER GRADE
Excellent appearance, a combination of First
and Second Grades.
Bundles 2 ft. and up; 2 ft. and 3 ft. bundles
up to 40% footage.
(NOTE: 5% 1¼ ft. bundles allowed in Second and Better
jointed flg. only.)

SECOND GRADE
Variegated appearance, varying sound wood
characteristics of species.
Bundles 2 ft. and up; 2 ft. and 3 ft. bundles
up to 45% footage.

THIRD AND BETTER GRADE
A combination of First, Second and Third Grades.
Bundles 1¼ ft. and up; 1¼ ft. to 3 ft. bundles as produced
up to 50% footage.

THIRD GRADE
Rustic appearance, all wood characteristics of species,
serviceable economical floor after filling.
Bundles 1¼ ft. and up; 1¼ ft. to 3 ft. bundles as produced
up to 65% footage.

HARDWOOD FLOORING GRADES *(cont.)*
Pecan Flooring

*FIRST GRADE RED (Special Order)
Same as First Grade except face all heartwood.

*FIRST GRADE WHITE (Special Order)
Same as First Grade except face all bright sapwood.

FIRST GRADE
Excellent appearance, natural color variation, limited character marks, unlimited sap.
Bundles 2 ft. and up; 2 ft. and 3 ft. bundles up to 25% footage.

*SECOND GRADE RED (Special Order Only)
Same as Second Grade except face all heartwood.

SECOND GRADE
Variegated appearance, varying sound wood characteristics of species.
Bundles 1¼ ft. and up; 1¼ ft. to 3 ft. bundles as produced up to 40% footage.

THIRD GRADE
Rustic appearance, all wood characteristics of species, a serviceable economical floor after filling.
Bundles 1¼ ft. and up; 1¼ ft. to 3 ft. bundles as produced up to 60% footage.

Prefinished Oak Flooring

*PRIME (Special Order Only)
Excellent appearance, natural color variation, limited
character marks, unlimited sap.
Bundles 1¼ ft. and up; average length 3½ ft.

STANDARD & BETTER GRADE
Combination of Standard and Prime.
Bundles 1¼ ft. and up; average length 3 ft.

STANDARD GRADE
Variegated appearance, varying sound wood
characteristics of species, a sound floor.
Bundles 1¼ ft. and up; average length 2¾ ft.

*TAVERN & BETTER GRADE (Special Order Only)
Combination of Prime, Standard and Tavern; all wood
characteristics of species.
Bundles 1¼ ft. and up; average length 3 ft.

TAVERN GRADE
Rustic appearance, all wood characteristics of species,
a serviceable economical floor.
Bundles 1¼ ft. and up; average length 2¼ ft.

---

*1¼ ft. Shorts (Red & White may be mixed). Unique Variegated
Appearance. Lengths 9 inches to 18 inches.
Bundles average nominal 1¼ ft. Production limited.

*No. 1 Common and Better Shorts. A combination grade,
Clear, Select, and No. 1 Common, 9 inches to 18 inches.

*No. 2 Common Shorts. Same as No. 2 Common,
except length 9 inches to 18 inches.

**Quarter Sawn — Special Order Only.

HARDWOOD FLOORING GRADES, SIZES, COUNTS AND WEIGHTS

"Nominal" is used by the *trade*, but it is not always the actual size, which may be $\frac{1}{32}$" less than the so-called nominal size. "Actual" is the *mill* size for thickness and face width, excluding tongue width. "Counted" size determines the board feet in a shipment. Pieces less than 1 inch in thickness are considered to be 1".

Nominal	Actual	Counted	Weights M Ft.
Tongue and Groove-end Matched			
**$\frac{3}{4}$ x $3\frac{1}{4}$"	$\frac{3}{4}$ x $3\frac{1}{4}$"	1 x 4"	2210 lbs.
$\frac{3}{4}$ x $2\frac{1}{4}$"	$\frac{3}{4}$ x $2\frac{1}{4}$"	1 x 3"	2020 lbs.
$\frac{3}{4}$ x 2"	$\frac{3}{4}$ x 2"	1 x $2\frac{3}{4}$"	1920 lbs.
$\frac{3}{4}$ x $1\frac{1}{2}$"	$\frac{3}{4}$ x $1\frac{1}{2}$"	1 x $2\frac{1}{4}$"	1820 lbs.
**$\frac{3}{8}$ x 2"	$\frac{11}{32}$ x 2"	1 x $2\frac{1}{2}$"	1000 lbs.
**$\frac{3}{8}$ x $1\frac{1}{2}$"	$\frac{11}{32}$ x $1\frac{1}{2}$"	1 x 2"	1000 lbs.
**$\frac{1}{2}$ x 2"	$\frac{15}{32}$ x 2"	1 x $2\frac{1}{2}$"	1350 lbs.
**$\frac{1}{2}$ x $1\frac{1}{2}$"	$\frac{15}{32}$ x $1\frac{1}{2}$"	1 x 2"	1300 lbs.
Square Edge			
**$\frac{5}{16}$ x 2"	$\frac{5}{16}$ x 2"	face count	1200 lbs.
**$\frac{5}{16}$ x $1\frac{1}{2}$"	$\frac{5}{16}$ x $1\frac{1}{2}$"	face count	1200 lbs.
Special Thickness (T and G, End Matched)			
**$\frac{33}{32}$ x $3\frac{1}{4}$"	$\frac{33}{32}$ x $3\frac{1}{4}$"	$\frac{5}{4}$ x 4"	2400 lbs.
**$\frac{33}{32}$ x $2\frac{1}{4}$"	$\frac{33}{32}$ x $2\frac{1}{4}$"	$\frac{5}{4}$ x 3"	2250 lbs.
**$\frac{33}{32}$ x 2"	$\frac{33}{32}$ x 2"	$\frac{5}{4}$ x $2\frac{3}{4}$"	2250 lbs.
Jointed Flooring — i.e., Square Edge			
**$\frac{3}{4}$ x $2\frac{1}{2}$"	$\frac{3}{4}$ x $2\frac{1}{2}$"	1 x $3\frac{1}{4}$"	2160 lbs.
**$\frac{3}{4}$ x $3\frac{1}{4}$"	$\frac{3}{4}$ x $3\frac{1}{4}$"	1 x 4"	2300 lbs.
**$\frac{3}{4}$ x $3\frac{1}{2}$"	$\frac{3}{4}$ x $3\frac{1}{2}$"	1 x $4\frac{1}{4}$"	2400 lbs.
**$\frac{33}{32}$ x $2\frac{1}{2}$"	$\frac{33}{32}$ x $2\frac{1}{2}$"	$\frac{5}{4}$ x $3\frac{1}{4}$"	2500 lbs.
**$\frac{33}{32}$ x $3\frac{1}{2}$"	$\frac{33}{32}$ x $3\frac{1}{2}$"	$\frac{5}{4}$ x $4\frac{1}{4}$"	2600 lbs.

**Special Order Only

NAIL CHART FOR APPLICATION OF STRIP FLOORING

Flooring Nominal Size, Inches	Size of Fasteners	Spacing of Fasteners
Tongue and Groove Flooring Must Be Blind Nailed		
¾ x 1½" ¾ x 2¼" ¾ x 3¼" ¾ x 3" to 8" plank**	2" machine driven fasteners; 7d or 8d screw or cut nail	10" – 12" apart* 8" apart into and between joists
Following Flooring Must Be Laid on a Subfloor		
½ x 1½" ½ x 2"	1½" machine driven fastener; 5d screw, cut steel or wire casting nail	10" apart
⅜ x 1½" ⅜ x 2"	1¼" machine driven fastener, or 4d bright wire casing nail	8" apart
Square-Edge Flooring as Follows, Face-Nailed — Through Top Face		
⁵⁄₁₆ x 1½" ⁵⁄₁₆ x 2"	1", 15-gauge fully barbed flooring brad	2 nails every 7"
⁵⁄₁₆ x 1⅓"	1", 15-gauge fully barbed flooring brad	1 nail every 5" on alternate sides of strip

*If subfloor is ½" plywood, fasten into each joist, with additional fastening between joists.
**Plank flooring over 4" wide must be installed over a subfloor.

CUTAWAY VIEW SHOWING VARIOUS LAYERS OF MATERIAL UNDER STRIP FLOORING

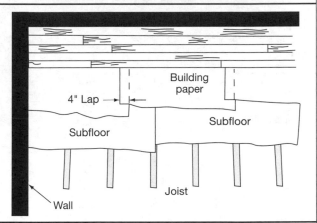

ADHESIVE FOR RESILIENT FLOOR TILE

Type and Use	Approximate Coverage in Square Feet per Gallon
Primer – For treating on or below grade concrete subfloors before installing asphalt tile.	250 to 350
Asphalt cement – For installing asphalt tile over primed concrete subfloors in direct contact with the ground.	200
Emulsion adhesive – For installing asphalt tile over lining felt.	130 to 150
Lining paste – For cementing lining felt to wood subfloor.	160
Floor and wall size – For priming chalky or dusty suspended concrete subfloors before installing resilient tile other than asphalt.	200 to 300
Waterproof cement – Recommended for installing linoleum tile, rubber and cork tile over any type of suspended subfloor in areas where surface moisture is a problem.	130 to 150

ESTIMATING MATERIAL FOR FLOOR TILE

Number of Tiles

Square Footage of Floor	9" x 9"	12" x 12"	6" x 6"	9" x 18"
1	2	1	4	1
2	4	2	8	2
3	6	3	12	3
4	8	4	16	4
5	9	5	20	5
6	11	6	24	6
7	13	7	28	7
8	15	8	32	8
9	16	9	36	8
10	18	10	40	9
20	36	20	80	18
30	54	30	120	27
40	72	40	160	36
50	89	50	200	45
60	107	60	240	54
70	125	70	280	63
80	143	80	320	72
90	160	90	360	80
100	178	100	400	90
200	356	200	800	178
300	534	300	1200	267
400	712	400	1600	356
500	890	500	2000	445

Note: Add 5% for waste.

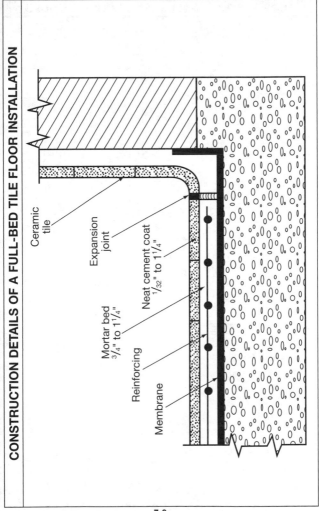

CONSTRUCTION DETAILS OF A FULL-BED TILE FLOOR INSTALLATION

Ceramic tile

Expansion joint

Neat cement coat 1/32" to 1/4"

Mortar bed 3/4" to 1 1/4"

Reinforcing

Membrane

7-9

THIN-BED TILE WALL INSTALLATION DETAILS

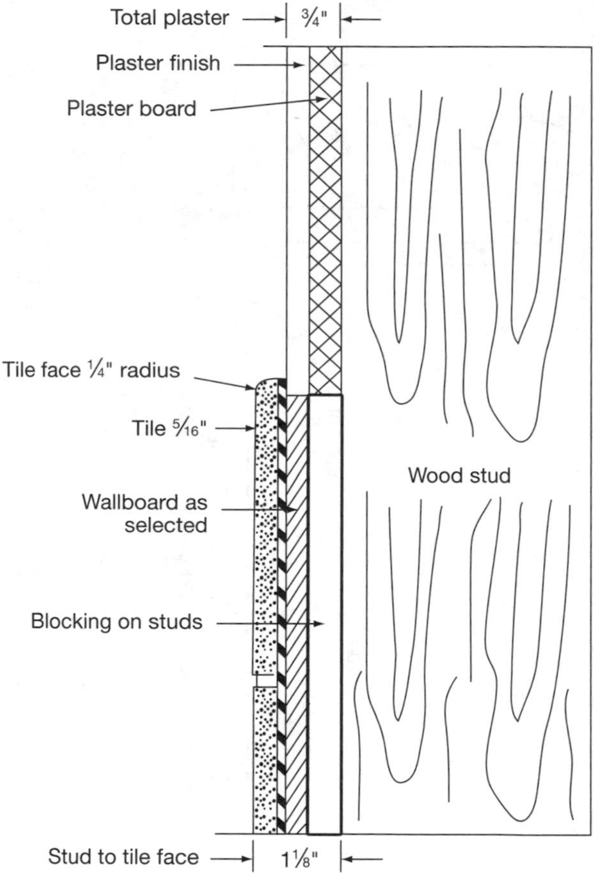

Total plaster — 3/4"

Plaster finish

Plaster board

Tile face 1/4" radius

Tile 5/16"

Wallboard as selected

Blocking on studs

Wood stud

Stud to tile face — 1 1/8"

7-10

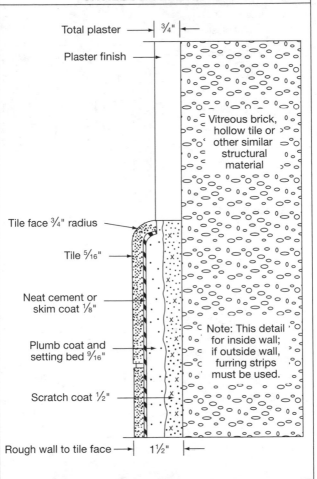

FULL-MORTAR-BED TILE DETAILS OVER A VITREOUS SOLID WALL

Total plaster — 3/4"

Plaster finish —

Vitreous brick, hollow tile or other similar structural material

Tile face 3/4" radius

Tile 5/16"

Neat cement or skim coat 1/8"

Plumb coat and setting bed 9/16"

Note: This detail for inside wall; if outside wall, furring strips must be used.

Scratch coat 1/2"

Rough wall to tile face — 1 1/2"

7-11

FULL-MORTAR-BED ON WOOD STUDS FOR WALL TILE

Total plaster — ¾"

Plaster finish

Plasterboard

Tile face ¾" radius

Tile ⁵⁄₁₆"

Neat cement or skim coat ⅛"

Plumb coat and setting bed ⁹⁄₁₆"

Scratch coat ½"

Metal lath

Cover wood studs with waterproof paper, or coat all faces of studs with asphaltum paint

Stud to tile face — 1½"

Wood stud

7-12

APPROXIMATE PAINT REQUIREMENTS FOR INTERIORS AND EXTERIORS

Distance Around the Room	Ceiling Height 8 Feet	Ceiling Height 8½ Feet	Ceiling Height 9 Feet	Ceiling Height 9½ Feet	Paint for Ceiling	Finish for Floors	For Each Door or Window
30 feet	⅝ gallon	⅝ gallon	¼ gallon	¾ gallon	1 pint	1 pint	
35 feet	¾ gallon	¾ gallon	¼ gallon	⅞ gallon	1 quart	1 pint	Each window and frame requires ¼ pint
40 feet	⅞ gallon	⅞ gallon	⅞ gallon	1 gallon	1 quart	1 quart	
45 feet	⅞ gallon	1 gallon	1 gallon	1⅛ gallons	3 pints	1 quart	
50 feet	1 gallon	1⅛ gallons	1⅛ gallons	1¼ gallons	3 pints	1 quart	
55 feet	1⅛ gallons	1⅛ gallons	1¼ gallons	1¼ gallons	2 quarts	3 pints	Each door and frame requires ½ pint
60 feet	1¼ gallons	1¼ gallons	1⅜ gallons	1⅜ gallons	2 quarts	3 pints	
70 feet	1⅜ gallons	1½ gallons	1½ gallons	1⅝ gallons	3 quarts	2 quarts	
80 feet	1½ gallons	1⅝ gallons	1¼ gallons	1⅞ gallons	1 gallon	5 pints	

APPROXIMATE PAINT REQUIREMENTS FOR INTERIORS AND EXTERIORS (cont.)

Distance Around the House	Average Height 12 Feet	Average Height 15 Feet	Average Height 18 Feet	Average Height 21 Feet	Average Height 24 Feet
60 feet	1 gallon	1¼ gallons	1½ gallons	1¼ gallons	2 gallons
76 feet	1¼ gallons	1½ gallons	2 gallons	2¼ gallons	2½ gallons
92 feet	1½ gallons	2 gallons	2½ gallons	2¾ gallons	3 gallons
108 feet	1¾ gallons	2¼ gallons	2¾ gallons	3¼ gallons	3¾ gallons
124 feet	2 gallons	2½ gallons	3¼ gallons	3¾ gallons	4¼ gallons
140 feet	2½ gallons	3 gallons	3½ gallons	4 gallons	4½ gallons
156 feet	2¾ gallons	3¼ gallons	4 gallons	4½ gallons	5¼ gallons
172 feet	3 gallons	3¼ gallons	4½ gallons	5 gallons	5¼ gallons

On interior work, for rough, sand-finished walls or unpainted gypsum board, add 50% to quantities; for each door or window, deduct ½ pint of materials for walls. For trim, add ⅛ to ⅕ of the amount required for the body. For exterior blinds, ½ gallon will cover 12 to 14 blinds, one coat.

DRYING TIMES OF COATINGS

Material	Touch	Recoat	Rub
Lacquer	1-10 min.	1½-3 hrs.	16-24 hrs.
Lacquer sealer	1-10 min.	30-45 min.	1 hr. (sand)
Paste wood filler	—	24-48 hrs.	—
Paste wood filler (Q.D.)	—	3-4 hrs.	—
Water stain	1 hr.	12 hrs.	—
Oil stain	1 hr.	24 hrs.	—
Spirit stain	zero	10 min.	—
Shading stain	zero	zero	
Non-grain raising stain	15 min.	3 hrs.	—
NGR stain (quick-dry)	2 min.	15 min.	—
Pigment oil stain	1 hr.	12 hrs.	—
Pigment oil stain (Q.D.)	1 hr.	3 hrs.	—
Shellac	15 min.	2 hrs.	12-18 hrs.
Shellac (wash coat)	2 min.	30 min.	—
Varnish	1½ hrs.	18-24 hrs.	24-48 hrs.
Varnish (Q.D. synthetic)	½ hr.	4 hrs.	12-24 hrs.

Note: Average times. Different products will vary.

EXTERIOR PAINT SELECTION CHART

Surface	Aluminum	Cement Base Paint	Exterior Clear Finish	House Paint	Metal Roof Paint	Porch-and-Deck Paint	Primer or Undercoater	Rubber Base Paint	Spar Varnish	Transparent Sealer	Trim-and-Trellis Paint	Wood Stain	Metal Primer
Wood													
Natural finish	—	—	P	—	—	—	—	—	P	P	—	P	—
Porch floor	—	—	—	—	—	P+	—	—	—	—	—	—	—
Shingle roof	—	—	—	P+	—	—	—	—	—	—	—	P+	—
Shutters and trim	—	—	—	P+	—	—	—	—	—	—	P	—	—
Siding	P	P	—	P+	—	—	P	—	—	—	—	P+	—
Windows	P	P	—	P+	—	—	—	—	—	—	P	P+	—
Masonry													
Asbestos cement	P	—	—	P+	—	—	—	P	—	—	—	—	—
Brick	P	P	P	P+	—	—	—	P	—	P	—	—	—
Cement & cinder block	—	P	P	P+	—	—	—	P	—	P	—	—	—
Cement porch floor	—	P+	—	—	—	P+	P	P	—	—	—	—	—
Stucco	P	P	P	P+	—	—	—	P	—	P	—	—	—
Metal													
Copper	P+	—	P	—	—	—	—	—	P	P	—	—	—
Galvanized	P+	—	P	P+	P+	—	—	P	—	—	—	—	P
Iron	—	—	—	—	—	—	—	—	—	—	—	—	P
Roofing	P+	—	—	—	P+	P+	—	—	—	—	—	—	P
Siding	P	—	—	P+	—	—	—	—	—	—	—	—	P
Windows, aluminum	P+	—	—	P+	—	—	—	—	—	—	—	—	P
Windows, steel	P+	—	—	P+	—	—	—	—	—	—	—	—	P

COMPARISON OF PAINT BINDERS' PRINCIPAL PROPERTIES

	Alkyd	Cement	Epoxy	Latex	Oil	Phenolic	Rubber	Moisture Curing Urethane	Vinyl
Ready for use	Yes	No	No[3]	Yes	Yes	Yes	Yes	Yes	Yes
Brushability	A	A	A	+	+	A	A	A	—
Odor	+[1]	+	—	+	A	A	A	—	+
Cure normal temp.	A	A	A	+	—	A	+	+	+
Cure low temp.	A	A	—	A	—	A	A	+	+
Film build/coat	A	+	+	+	+	A	A	+	—
Safety	A	+	A	A	A	A	—	—	—
Use on wood	A	—	A	+	A	A	+	A	—
Use on fresh conc.	—	+	+	+	—	—	+	A	+
Use on metal	+	—	+	—	+	+	A	A	+
Corrosive service	A	—	+	—	—	A	A	A	+
Gloss – choice	+	—	+	—	A	+	A	A	A
Gloss – retention	+	X	—	X	—	+	A	A	+
Color – initial	+	A	A	+	A	—	+	+	+
Color – retention	+	—	A	+	A	—	+	—	A
Hardness	A	+	+	A	—	+	+	+	—
Adhesion	A	A	+	A	+	A	A	+	A
Flexibility	A	—	+	+	+	A	A	+	+
Resistance to:									
Abrasion	A	A	+	A	—	+	A	+	+
Water	A	A	A	A	A	+	+	+	+
Acid	A	—	A	A	—	A	+	+	+
Alkali	A	+	+	A	—	A	+	+	+
Strong solvent	—	+	+	A	—	A	—	+	A
Heat	A	+	A	+	A	A	+[2]	A	—
Moisture permeability	Mod.	V. High	Low	High	Mod.	Low	Low	Low	Low

+ = Among the best for this property — = Among the poorest for this property A = Average X = Not applicable [1] Odorless type [2] Special types [3] Two component type

SINGLE-ROLL WALLPAPER REQUIREMENTS

Size of room	Height of ceiling 8'	9'	10'	Yards of border	Rolls for ceiling
4 x 8	6	7	8	9	2
4 x 10	7	8	9	11	2
4 x 12	8	9	10	12	2
6 x 10	8	9	10	12	2
6 x 12	9	10	11	13	3
8 x 12	10	11	13	15	4
8 x 14	11	12	14	16	4
10 x 14	12	14	15	18	5
10 x 16	13	15	16	19	6
12 x 16	14	16	17	20	7
12 x 18	15	17	19	22	8
14 x 18	16	18	20	23	8
14 x 22	18	20	22	26	10
15 x 16	15	17	19	23	8
15 x 18	16	18	20	24	9
15 x 20	17	20	22	25	10
15 x 23	19	21	23	28	11
16 x 18	17	19	21	25	10
16 x 20	18	20	22	26	10
16 x 22	19	21	23	28	11
16 x 24	20	22	25	29	12
16 x 26	21	23	26	31	13
17 x 22	19	22	24	23	12
17 x 25	21	23	26	31	13
17 x 28	22	25	28	32	15
17 x 32	24	27	30	35	17
17 x 35	26	29	32	37	18
18 x 22	20	22	25	29	12
18 x 25	21	24	27	31	14
18 x 28	23	26	28	33	16

This chart assumes use of the standard roll of wallpaper, eight yards long and 18" wide. Deduct one roll of side wallpaper for every two doors or windows of ordinary dimensions, or for each 50 square feet of opening.

CHAPTER 9
Plan Symbols

ARCHITECTURAL SYMBOLS	
Symbol	**Definition**
	Wall section No. 2 can be seen on drawing No. A-4.
	Detail section No. 3 can be seen on drawing No. A-5.
	Building section A-A can be seen on drawing No. A-6.
————————	Main object line
— — — —	Hidden or invisible line
—————— · —	Indicates center line
3" 3' 4"	Dimension lines
↓ ↓	Extension lines
⅌	Symbol indicates center line
///////////	Indicates wall suface
N	Indicates north direction

9-1

ARCHITECTURAL SYMBOLS *(cont.)*

Symbol	Definition
(ca) — – – —	Column line grid
⟨5⟩— or ⟩5⟩—	Partition type
⟨A⟩	Window type
(05)	Door number
05	Room number
(10'-0")	Ceiling height
△2	Revision marker
—⟩ ⟨—	Break in a continuous line
(3)—→	Refer to note #3
◑ 100'-0" — – –	Elevation marker
◇ 1 / A-5 2 / 3	Interior elevations 1,2,3 & 4 can be seen on drawing A-5. Direction of triangle indicates elevation.

PLUMBING PIPING

Description	Symbol	Description	Symbol
Soil, waste or leader (above grade)	————	Soft cold water	——SW——
Soil, waste or leader (below grade)	— — — —	Industrialized cold water	——ICW——
Vent	– – – –	Chilled drinking water supply	——DWS—
Combination waste and vent	——SV——	Chilled drinking water return	——DWR——
Acid waste	——AW——	Hot water	– ·· – ·· –
Acid vent	– – AV – · –	Hot water return	– – – – –
Indirect drain	——IW——	Sanitizing hot water supply (180F)	–/– – –/
Storm drain	—— S ——	Sanitizing hot water return (180F)	–/– – –/
Cold water	– ·· – ·· –	Industrialized hot water supply	——IHW——

9-3

PLUMBING PIPING (cont.)

Industrialized hot water return	—— IHR ——	Gas – low pressure	— G — G —
Tempered water supply	—— TWS ——	Gas – medium pressure	—— MG ——
Tempered water return	—— TWR ——	Gas – high pressure	—— HG ——
Fire line	—— F —— F ——	Compressed air	—— A ——
Wet standpipe	—— WSP ——	Vacuum	—— V ——
Dry standpipe	——DSP——	Vacuum cleaning	——VC——
Combination standpipe	——CSP——	Oxygen	—— O ——
Main supplies sprinkler	—— S ——	Liquid oxygen	—— LOX ——
Branch and head sprinkler	——o——o——	Nitrogen	—— N ——

Liquid nitrogen	——LN——	
Nitrous oxide	——NO——	
Hydrogen	——H——	
Helium	——HE——	
Argon	——AR——	
Liquid petroleum gas	——LPG——	
Industrial waste	——INW——	

Pneumatic tubes tube runs	——PN——	
Cast iron	——CI——	
Culvert pipe	——CP——	
Clay tile	——CT——	
Ductile iron	——DI——	
Reinforced concrete	——RCP——	
Drain – open tile or agricultural tile	═══ ═══ ═══	

HEATING PIPING

Description	Symbol	Description	Symbol
High pressure steam	——HPS——	Make up water	——MU——
Medium pressure steam	——MPS——	Air relief line	—— V ——
Low pressure steam	——LPS——	Fuel oil suction	——FOS——
High pressure return	——HPR——	Fuel oil return	——FOR——
Medium pressure return	——MPR——	Fuel oil vent	——FOV——
Low pressure return	——LPR——	Compressed air	—— A ——
Boiler blow off	—— BD ——	Hot water heating supply	——HW——
Condensate or vacuum pump discharge	——VPD——	Hot water heating return	——HW——
Feedwater pump discharge	——PPD——		

9-6

AIR CONDITIONING PIPING

Refrigerant liquid	——— RL ———	Chilled water return	———CHWR———
Refrigerant discharge	——— RD ———	Make up water	——— MU ———
Refrigerant suction	——— RS ———	Humidification line	——— H ———
Condenser water supply	———CWS———	Drain	——— D ———
Condenser water return	———CWR———	Brine supply	——— B ———
Chilled water supply	———CHWS———	Brine return	——— BR ———

PIPING SYMBOLS

Valves, Fittings and Specialties

Gate		Concentric reducer		
Globe		Eccentric reducer		
Check		Pipe guide		
Butterfly		Pipe anchor		
Solenoid		Flow direction		
Lock shield		Elbow looking up		
2-Way automatic control		Elbow looking down		
3-Way automatic control		Pipe pitch up or down	Up/Down	
Gas cock		Expansion joint		
Plug cock		Expansion loop		
Flanged joint		Flexible connection		
Union				
Cap		Thermostat		
Strainer		Thermostatic trap		

PIPING SYMBOLS (cont.)

Float and thermostatic trap	F&T	Hose bibb	
Thermometer		Elbow	
Pressure gauge		Tee	
		'Y'	
Flow switch	FS	OS & Y gate	
Pressure switch	P	Shock absorber	
Pressure reducing valve		House trap	
Temperature and pressure relief valve		'P' trap	
Humidistat	H	Floor drain	
Aquastat	A	Indirect waste	—— IW ——
Air vent		Sanitary below grade	— — S — —
Meter	M	Sanitary above grade	—— S ——

Storm below grade	— — —ST— — —	Gas-low pressure	———G———
Storm above grade	———ST———	Gas-medium pressure	———MG———
Vent	— — — — — —	Gas-high pressure	———HG———
Combination waste & vent	———CWV———	Compressed air	———CA———
Acid waste below grade	— — —AW— — —	Vacuum	———V———
Acid waste above grade	———AW———	Vacuum cleaning	———VC———
Acid vent	— — —AV— — —	Nitrogen	———N———
Cold water	——— — —CW	Nitrous oxide	———N_2O———
Hot water	——— — —HW	Oxygen	———O———
Hot water circulation	——— — — —HWC	Liquid oxygen	———LOX———
Drinking water supply	———DWS———	Liquid petroleum gas	———LPG———
Drinking water return	———DWR———		

FIRE PROTECTION PIPING SYMBOLS

Fire protection water supply ——— F ———

Wall fire department connection

Wet standpipe ——— WSP ———

Dry standpipe ——— DSP ———

Sidewalk fire department connection

Combination standpipe ——— CSP ———

Fire hose rock   FHR

Automatic fire sprinkler ——— SP ———

Upright fire sprinkler heads

Surface mounted fire hose cabinet   FHC

Pendent fire sprinkler heads

Recessed fire hose cabinet   FHC

Fire hydrant

PLUMBING FIXTURE SYMBOLS

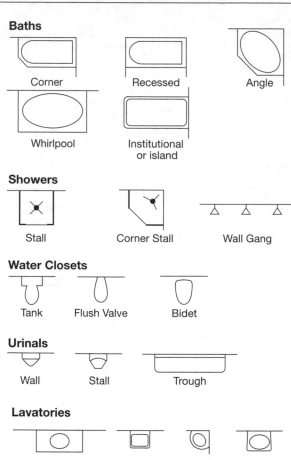

Baths

Corner

Recessed

Angle

Whirlpool

Institutional
or island

Showers

Stall

Corner Stall

Wall Gang

Water Closets

Tank

Flush Valve

Bidet

Urinals

Wall

Stall

Trough

Lavatories

Vanity

Wall

Counter

Pedestal

PLUMBING FIXTURE SYMBOLS *(cont.)*

Kitchen Sinks

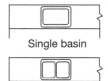

Single basin

Single drainboard

Twin basin

Double drainboard

Drinking Fountains or Electric Water Coolers

Floor or wall   Recessed   Semirecessed

Dishwasher

Laundry Trays

Single   Double

Service Sinks

Wall   Floor

Wash Fountains

Circular   Semicircular

Hot Water

Heater   Tank

Separators

Gas   Oil

MATERIAL INDICATION SYMBOLS

Material	Plan	Elevation	Section
Wood	Floor areas left blank	Siding Panel	Framing Finish
Brick	Face Common	Face or common	Same as plan view
Stone	Cut Rubble	Cut Rubble	Cut Rubble

9-14

Material			
Concrete	(symbol)	(symbol)	Same as plan view
Concrete block	(symbol)	(symbol)	Same as plan view
Earth	None	None	(symbol)
Glass	—	(symbol)	Large scale / Small scale
Insulation	Same as section	Insulation	Loose fill or batt / Board

9-15

MATERIAL INDICATION SYMBOLS *(cont.)*

Material	Plan	Elevation	Section
Plaster	Same as section	Plaster	Stud / Lath and plaster
Structural steel	— — — / ————	Indicate by note	
Sheet metal flashing	Indicate by note		Show contour

Material	Floor	Wall	Symbol
Tile			
Porous fill	None	None	
Plywood	Indicated by note	Indicated by note	

9-17

MATERIAL INDICATION SYMBOLS (cont.)

Material	Plan	Elevation	Section
Batt insulation		None	Same as plan
Rigid insulation		None	Same as plan
Glass			Small scale Large scale

9-18

Gypsum wallboard			Same as plan
Acoustical		None	
Ceramic wall tile			Same as plan
Floor tile		None	

9-19

LANDSCAPE SYSTEMS AND GRAPHICS

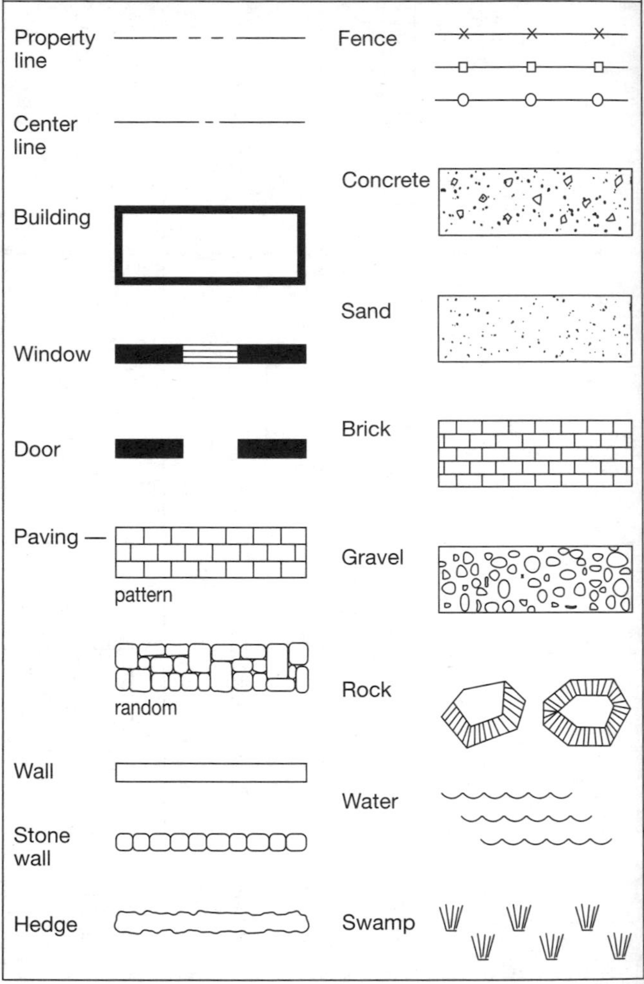

Property line

Center line

Building

Window

Door

Paving —

pattern

random

Wall

Stone wall

Hedge

Fence

Concrete

Sand

Brick

Gravel

Rock

Water

Swamp

9-20

LANDSCAPE SYSTEMS AND GRAPHICS *(cont.)*

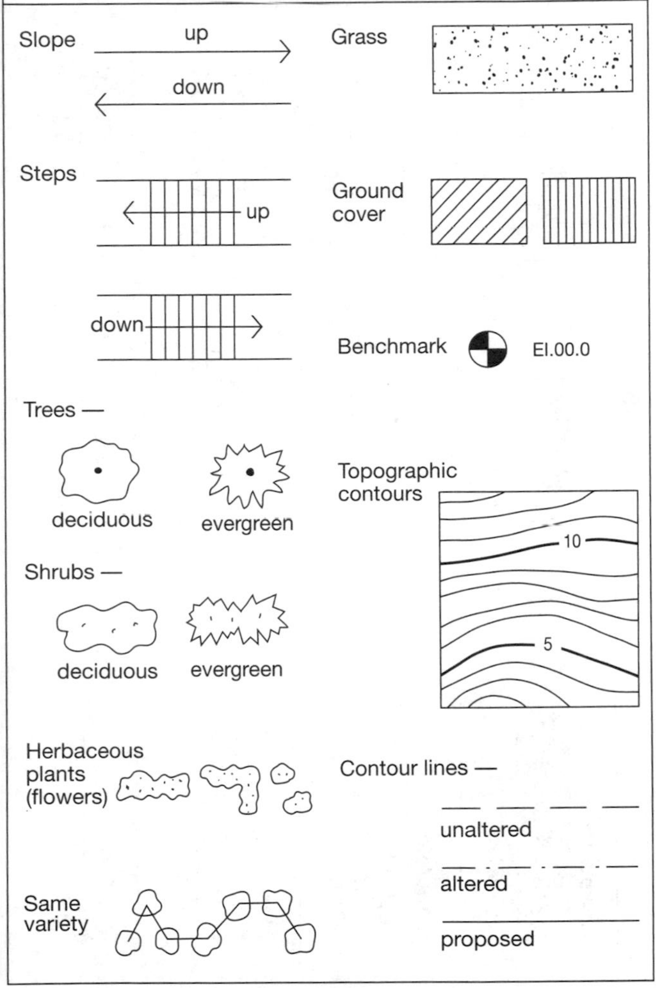

Slope   up →   ← down

Grass

Steps   ← up   down →

Ground cover

Benchmark   El.00.0

Trees —

deciduous   evergreen

Topographic contours

10

5

Shrubs —

deciduous   evergreen

Herbaceous plants (flowers)

Contour lines —

_ _ _ _ _ _
unaltered

_ . _ . _ . _
altered

_____
proposed

Same variety

MATERIALS SYMBOLS

Earthworks

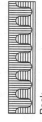
Earth/compact fill

Porous fill/gravel

Rock

Concrete

Cast-in-place/precast

Lightweight

Sand/mortar/plaster/cut stone

Masonry

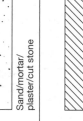
Adobe/rammed earth

Common/face

Fire brick

Concrete block

Gypsum block

Structural facing tile

Stone

Bluestone/slate/
soapstone/flagging

Rubble

Marble

Metal

Aluminum

Brass/bronze

Steel/other metals

Wood

Finish

Rough

Blocking

Hardboard

Plywood – large scale

Plywood – small scale

9-23

MATERIALS SYMBOLS (cont.)

Glass

Glass

Structural

Glass block

Insulation

Batt/loose fill

Rigid

Spray/foam

Finishes

Acoustical tile

Ceramic tile – large scale

Ceramic tile – small scale

Carpet and pad

Gypsum wallboard

Metal lath and plaster

9-24

Finishes (cont.)

Plastic

Resilient flooring/plastic laminate

Terrazzo

Plan and Section Indications
Partition Indications

Wood stud

Metal stud

Special finish face

Elevation Indications

Brick

Ceramic tile

Concrete/plaster

Glass

Sheet metal

Shingles/siding

9-25

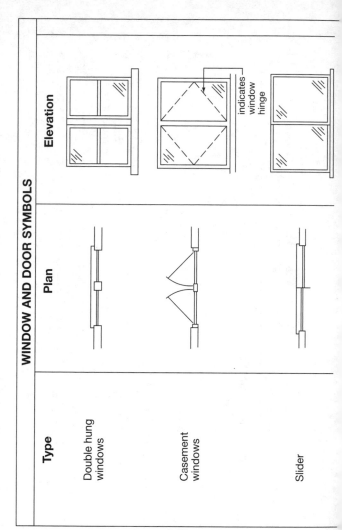

WINDOW AND DOOR SYMBOLS

Type	Plan	Elevation
Double hung windows		
Casement windows		indicates window hinge
Slider		

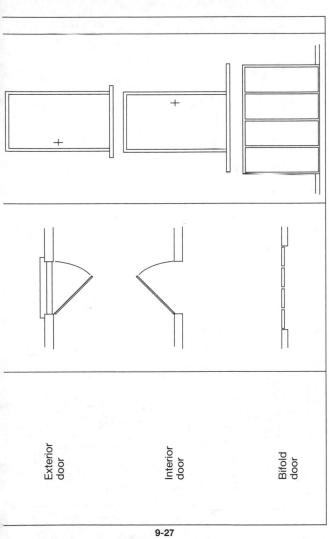

Exterior
door

Interior
door

Bifold
door

9-27

WINDOW AND DOOR SYMBOLS *(cont.)*

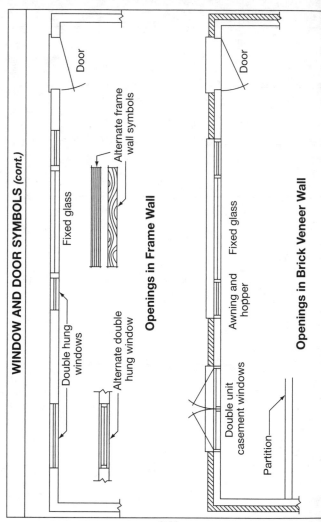

Openings in Frame Wall

Door

Fixed glass

Alternate frame wall symbols

Double hung windows

Alternate double hung window

Openings in Brick Veneer Wall

Door

Fixed glass

Awning and hopper

Double unit casement windows

Partition

9-28

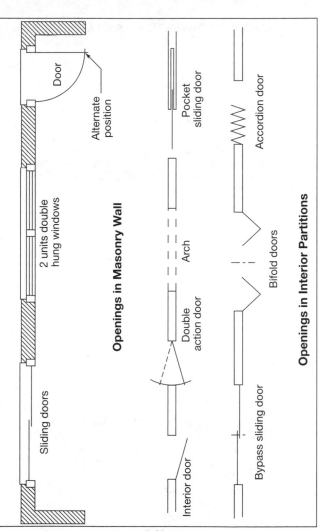

Openings in Masonry Wall

Sliding doors

2 units double hung windows

Door

Alternate position

Openings in Interior Partitions

Interior door

Double action door

Arch

Pocket sliding door

Bypass sliding door

Bifold doors

Accordion door

9-29

Graphic Symbols

The symbols shown are those that seem to be the most common and acceptable, judged by the frequency of use by the architectural offices surveyed. This list can and should be expanded by each office to include symbols generally used by it, but not indicated here. Adoption of these symbols as standard practice is desirable to improve communication in the industry.

 Stair direction symbol

 North point
to be placed on each
floor plan, generally in
lower right hand corner
of drawings

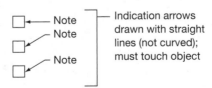 Indication arrows
drawn with straight
lines (not curved);
must touch object

Indicates section number

Indicates drawing sheet
on which section is shown

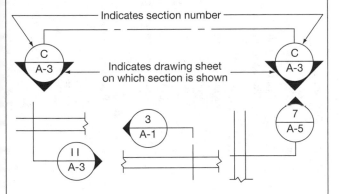

Section Lines and Section References

Indicates detail number

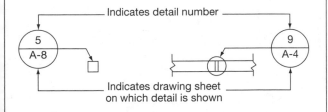

Indicates drawing sheet
on which detail is shown

Detail References

DRAWING CONVENTIONS AND SYMBOLS *(cont.)*

461.0' (in rectangle)	New or required point elevation
+ 461.0'	Existing point elevation (plan)
288 (dashed contour)	Existing contours elevation noted on high side
320 (solid contour)	New contours elevation noted on high side
● TB-1	Test boring

C / A-9	Building section, Reference drawing number
7 / A-11	Wall section or elevation, Reference drawing number
7 / A-12	Detail, Reference drawing number
1302	Room/space number
354	Equipment number

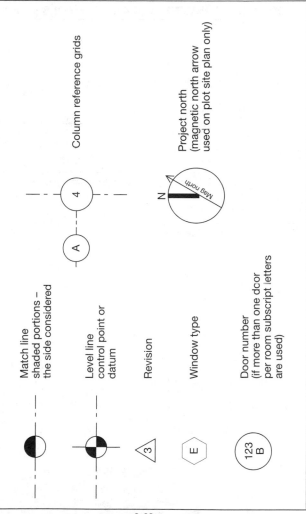

Match line
shaded portions –
the side considered

Level line
control point or
datum

Revision

Window type

Door number
(if more than one dcor
per room subscript letters
are used)

Column reference grids

Project north
(magnetic north arrow
used on plot site plan only)

Mag north

N

A

4

3

E

123
B

DRAWING CONVENTIONS AND SYMBOLS *(cont.)*

Dash and dot ⎯⎯⎯

Center lines, projections, existing elevations lines

Dash and double dot line ⎯⎯⎯

Property lines, boundary lines

Dotted line ⎯⎯⎯

Hidden, future or existing construction to be removed

Break line ⎯⎯⎯

To break off parts of drawing

Linework

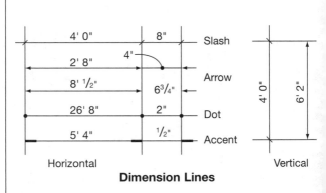

Horizontal

Vertical

Dimension Lines

CHAPTER 10
Glossary, Abbreviations and Symbols

GLOSSARY

A

ABS (acrylonitrile butadiene styrene) plastic pipe used for plumbing construction

abut joining end-to-end

accelerator a concrete additive used to speed the curing time of freshly poured concrete

acoustical referring to the study of sound transmission or reduction

adhesive a bonding material used to bond two materials together

adjacent touching; next to

aggregate fine, lightweight, coarse, or heavyweight grades of sand, vermiculite, perlite, or gravel added to cement or concrete or plaster

air-drying method of removing excess moisture from lumber using natural circulation of air

air handling unit a mechanical unit used for air conditioning or movement of air as in direct supply or exhaust of air within a structure

allowable load maximum supportable load of any construction components(s)

allowable span maximum length permissible for any framing component without support

anchor bolt a J- or L-shaped steel rod threaded on one end for securing structural members to concrete or masonry

apron a piece of window trim that is located beneath the window sill; also used to designate the front of a building such as the concrete apron in front of a garage

arbor an axle on which a cutting tool is mounted; it is a common term used in reference to the mounting of a circular saw blade

architect's scale a rule with scales indicating feet, inches, and fractions of inches

asphalt the general term for a black material produced as a by-product of oil (asphalt) or coal (pitch or coal tar)

asphalt shingle a composition-type shingle used on a roof and made of a saturated felt paper with ground-up pieces of stone embedded and held in place by asphaltum

asphalt shingles shingles made of asphalt or tar-impregnated paper with a mineral material embedded; very fire resistant

GLOSSARY *(cont.)*

awl a tool used to mark wood with a scratch mark; can be used to produce pilot holes for screws

awning window a window that is hinged at the top and the bottom swings outward

B

backfill any deleterious material (sand, gravel, and so on) used to fill an excavation

backhoe self-powered excavation equipment that digs by pulling a boom mounted bucket toward itself

backsplash the vertical part of a countertop that runs along the wall to prevent splashes from marring the wall

balloon framing wall construction extending from the foundation to the roof structure without interruption; used in residential construction only

baluster that part of the staircase which supports the handrail or bannister

balustrade a complete handrail assembly; includes the rails, the balusters, subrails, and fillets

bank plug piece of lumber (usually 2" x 4") driven into the ground to stand some distance, usually 24", above ground level; surveyors place nails in the bank plugs a given

distance above the road surface so a string line can be stretched between the plugs to measure grade

bannister that part of the staircase which fits on top of the balusters

baseboard molding covering the joint between a finished wall and the floor

base shoe a molding added at the bottom of a baseboard; used to cover the edge of finish flooring or carpeting

batt insulation an insulating material formed into sheets or rolls with a foil or paper backing; to be installed between framing members

batten a narrow piece of wood used to cover a joint

batter boards boards used to frame in the corners of a proposed building while the layout and excavating work takes place

beam a horizontal framing member; may be made of steel or wood; usually refers to a wooden beam at least 5 inches thick and at least 2 inches wider than it is thick

bearing partition an interior divider or wall that supports the weight of the structure above it

bearing wall a wall having weight-bearing properties associated with holding up a building's roof or second floor

benching making steplike cuts into a slope; used for erosion control or to tie a new fill into an existing slope

benchmark point of known elevation from which the surveyors can establish all their grades

berm a raised earth embankment; the shoulder of a paved road; the area between the curb and the gutter and a sidewalk

bevel a tool that can be adjusted to any angle; it helps make cuts at the number of degrees that is desired and is a good device for transferring angles from one place to another

bibb also know as a *hose bibb*; a faucet used to connect a hose

bi-fold a double-leaf door used primarily for closet doors in residential construction

bird mouth a notch cut into a roof rafter so that it can rest smoothly on the top plate

bitumen the general term used to identify asphalt and coal tar

blistering the condition that paint presents when air or moisture is trapped underneath and makes bubbles that break into flaky particles and ragged edges

blocking a piece of wood fastened between structural members to strengthen them; generally, solid or cross-tie wood or metal cross-tie members to perform the same task

board foot a unit of lumber measure equaling 144 cubic inches; the base unit (B.F.) is 1 inch thick and 12 inches square or $1 \times 12 \times 12 = 144$ cubic inches

bond in masonry, the interlocking system of brick or block to be installed

borrow site an area from which earth is taken for hauling to a jobsite which is short of earth needed to build an embankment

bottom or heel cut the cutout of the rafter end which rests against the plate; also called the *foot* or *seat cut*

bow a term used to indicate an upward warp along the length of a piece of lumber that is laid

bow window a window unit that projects from an exterior wall

GLOSSARY (cont.)

brace an inclined piece of lumber applied to a wall or to roof rafters to add strength

bridging used to keep joists from twisting or bending

builder's level a tripod-mounted device that uses optical sighting to make sure that a straight line is sighted and that the reference point is level

building codes rules and regulations which are formulated in a code by a local housing authority or governing body

building paper also called tar paper, roofing paper, and a number of other terms; paper having a black coating of asphalt for use in weatherproofing

building permits a series of permits that must be obtained for building; allows for inspections of the work and for placing the house on the tax roles

butt to meet edge to edge, such as in a joining of wooden edges

C

calcium chloride a concrete admixture used for accelerating the cure time

California Bearing Ratio (CBR) a system used for determining the bearing capacity of a foundation

carriage a notched stair frame

casement a type of window hinged to swing outward

casing the trim that goes on around the edge of a door opening; also the trim around a window

catch basin a complete drain box made in various depths and sizes; water drains into a pit, then from it through a pipe connected to the box

catch point another name for hinge point or top of shoulder

caulk any type of material used to seal walls, windows, and doors to keep out the weather

cement a material that, when combined with water, hardens due to chemical reaction; the basis for a concrete mix

cement plaster a mixture of gypsum, cement, hydrated lime, sand, and water, used primarily for exterior wall finish

cementitious able to harden like cement

center line the point on stakes or drawings which indicate the halfway point between two sides

chair small device used to support horizontal rebar prior to the concrete placement

chord top or bottom member of a truss

clear and grub to remove all vegetation, trees, concrete, or anything that will interfere with construction inside the limits of the project

cleat any strip of material attached to the surface of another material to strengthen, support, or secure a third material

collar tie horizontal framing member tying the raftering together above the plate line

common rafter a structural member that extends without interruption from the ridge to the plate line in a sloped roof structure

compactor a machine for compacting soil; can be pulled or self powered

computer-aided drafting (CAD) computer-aided design and drafting

concrete a mixture of sand, gravel, and cement in water

condensation the process by which moisture in the air becomes water or ice on a surface (such as a window) whose temperature is colder than the air's temperature

contour line solid or dashed lines showing the elevation of the earth on a project

convection transfer of heat through the movement of a liquid or gas

corner beads these are metal strips that prevent damage to drywall corners

cornice the area under the roof overhang; usually enclosed or boxed in

crawl space the area under a floor that is not fully excavated; only excavated sufficiently to allow one to crawl under it to get at the electrical or plumbing devices

cripple jack a jack rafter with a cut that fits in between a hip and a valley rafter

cripple rafter a cripple rafter is not as long as the regular rafter used to span a given area

cripple stud a short stud that fills out the position where the stud would have been located if a window, door, or some other opening had not been there

cross brace wood or metal diagonal bracing used to aid in structural support between joists and beams

crows foot a lath set by the grade setter with markings to indicate the final grade at a certain point

cup to warp across the grain

curtain wall inside walls are often called curtain walls; they do not carry loads from roof or floors above them

cutting plane line a heavy broken line with arrows, letters, and numbers at each end indicating the section view that is being identified

D

dado a rectangular groove cut into a board across the grain

dampproofing moisture protection; a surfacing used to coat and protect concrete and masonry from moisture penetration

datum point see benchmark; identification of the elevation above mean sea level

dead load the weight of a structure and all its fixed components

deck the part of a roof that covers the rafters

deformed bar steel reinforcement bar with ridges to prevent the bar from loosening during the concrete curing process

diagonal brace a wood or metal member placed diagonally over wood or metal framing to add rigidity at corners and at 25'0" feet of unbroken wall space

dimension line a line on a drawing with a measurement indicating length

double plate usually refers to the practice of using two pieces of dimensional lumber for support over the top section or wall section

double trimmer double joists used on the sides of openings; double trimmers are placed without regard to regular joist spacings for openings in the floor for stairs or chimneys

downspouts pipes connected to the gutter to conduct rainwater to the ground or sewer

drain tile usually made of plastic, generally 4 inches in diameter, with a number of small holes to allow water to drain into it; laid along the foundation footing to drain the seepage into a sump or storm sewer

drop siding drop siding has a special groove, or edge cut into it; the edge lets each board fit into the next board and makes the boards fit together to resist moisture and weather

ductwork a system of pipes used to pass heated air along to all parts of a house; also used to distribute cold air for summer air conditioning

GLOSSARY (cont.)

E

earthwork excavating and grading

easement a portion of land on or off a property which is set aside for utility installations

eave the lowest edge on a gable roof

eaves eaves are the overhang of a roof projecting over the walls

eaves trough a gutter

elevation an exterior or interior orthographic view of a structure identifying the design and the materials to be used

elevation numbers the vertical distance above or below sea level

embankment area being filled with earth

F

face the exposed side of a framing or masonry unit; a type of brick; also called *common*

fascia a flat board covering the ends of rafters on the cornice or eaves; eave troughs are usually mounted to the fascia board

feathering raking new asphalt to join smoothly with the existing asphalt

finish any material used to complete an installation that provides an aesthetic or finished appearance

firebrick a special type of brick that is not damaged by fire; used to line the firebox in a fireplace

fire stop/draft stop/fire blocking a framing member used to reduce the ability of a fire's spread

firewall/fire separation wall/fire division wall any wall that is installed for the purpose of preventing the spread of fire

flashing metal or plastic strips or sheets used for moisture protection in conjunction with other construction materials

flat in roofing, any roof structure up to a 3:12 slope

flue the passage through a chimney

flush to be even with

fly ash fine, powdery coal residue used with a hydraulic (water-resistant) concrete mix

footing the bottom-most member of a foundation; supports the full load of the structure above

form a temporary construction member used to hold permanent materials in place

foundation the base on which a house or building rests

GLOSSARY (cont.)

frostline the depth to which ground freezes in the winter

furring strips strips of wood attached to concrete or stone that form a nail base for wood or paneling

G

gable the simplest kind of roof; two large surfaces come together at a common edge forming an inverted V

galvanize a coating of zinc primarily used on sheet metal

galvanized iron sheet metal coated with zinc

gambrel roof a barn-shaped roof

gauge the thickness of metal or glass sheet material

girder a support for joists at one end; usually placed halfway between the outside walls and runs the length of the building

glaze to install glass

glu-lam (GLB) glue-laminated beam made from milled 2x lumber bonded together to form a beam

grade an existing or finished elevation in earthwork; a sloped portion of a roadway; sizing of gravel and sand; the structural classification of lumber

grade break a change in slope from one incline ratio to another

grade lath a piece of lath that the surveyor or grade setter has marked to indicate the correct grade to the operators

grade pins steel rods driven into the ground at each surveyor's hub; a string is stretched between them at the grade indicated on the survey stakes, or a constant distance above the grade

grader a power excavating machine with a central blade that can be angled to cast soil on either side; it has an independent hoist control on each side; also called a *blade*

gravel stop the edge metal used at the eaves of a built-up roof to hold the gravel on the roof

green uncured or set concrete or masonry; freshly cut lumber

grid system a system of metal strips that support a drop ceiling

ground-fault-circuit-inter-rupter (GFCI) or (GFI) an electrical receptacle installed for personal safety at outdoor locations and near water

guinea a survey marker driven to grade; it may be colored with paint or crayon; used for finishing and fine trimming; also called a *hub*

gusset a triangular or rectangular piece of wood or metal that is usually fastened to the joint of a truss to strengthen it; used primarily in making roof trusses

gutter a metal or wooden trough set below the eaves to catch and conduct water from rain and melting snow to a downspout

gypsum a chalk used to make wallboard; made into a paste, inserted between two layers of paper and allowed to dry; produces a plastered wall with certain fire-resisting characteristics

H

habitable space in residential construction, the interior areas of a residence used for eating, sleeping, living, and cooking; excludes bathrooms, storage rooms, utility rooms, and garages

hanger metal fabrication made for the purpose of placing and supporting joists and rafters

hardware any component used to hang, support, or position another component; e.g., door and window hardware, hangers, and so on

hardwood wood that comes from a tree that sheds it leaves

head joint the end face of a brick or concrete masonry unit to which the mortar is applied

header a framing member used to hide the ends of joists along the perimeter of a structure; also known as a *rim joist*; the horizontal structural framing member installed over wall openings to aid in the support of the structure above; also referred to as a *lintel*

header course in masonry, a horizontal row of brick laid perpendicular to the wall face; used to tie a double wythe brick wall together

hidden line a dashed line identifying portions of construction that are a part of the drawing but cannot be seen; e.g., footings on foundation plans or wall cabinetry in floor plans

hip rafters a member that extends diagonally from the corner of the plate to the ridge

hip roof a structural sloped roof design with sloped perimeters from ridge to plate line

hose bibb a faucet used to connect a hose

HVAC Heating, Ventilating and Air Conditioning. Term given to all heating and air conditioning systems; the mechanical portion of the CSI format, division 15

hydraulic cement a cement used in a concrete mix capable of curing under water

I

insulation any material capable of resisting thermal, sound, or electrical transmission

insulation resistance the R factor in insulation calculations

J

jack rafter a part of the roof structure raftering that does not extend the full length from the ridge beam to the top plate

jamb a jamb is the part that surrounds a door window frame; usually made of two vertical pieces and a horizontal piece over the top

joint compound material used with a paper of fiber tape for sealing indentations and breaks in drywall construction

joist a structural horizontal framing member used for floor and ceiling support systems

joist hangers metal brackets that hold up the joist; they are nailed to the girder, and the joist fits into the bracket

K

key a depression made in a footing so that the foundation or wall can be poured into the footing, preventing the wall or foundation from moving during changes in temperature or settling of the building

kicker blocks cement poured behind each bend or angle of water pipe for support; also called *thrust blocks*

kiln-dried lumber lumber that is seasoned under controlled conditions, removing from 6% to 12% of the moisture in green lumber

king stud a full-length stud from bottom plate to the top plate supporting both sides of a wall opening

knee wall vertical framing members supporting and shortening the span of the roof rafters

L

lateral underground electrical service

lath backup support for plaster; may be of wood, metal, or gypsum board

lavatory bathroom; vanity basin

lay-in ceiling a suspended ceiling system

leach line a perforated pipe used as a part of a septic system to allow liquid overflow to dissipate into the soil

ledger structural framing member used to support ceiling and roof joists at the perimeter walls

level-transit an optical device that is a combination of a level and a means for checking vertical and horizontal angles

lift any layer of material or soil placed upon another

live load any movable equipment or personal weight to which a structure is subjected

load the weight of a building

load conditions the conditions under which a roof must perform

lockset the doorknob and associated locking parts inserted in a door

M

masonry manufactured materials of clay (brick), concrete (CMU), and stone

mastic an adhesive used to hold tiles in place; also refers to adhesives used to glue many types of materials in the building process

mat asphalt as it comes out of a spreader box or paving machine in a smooth, flat form

maximum density and optimum moisture the highest point on the moisture density curve; considered the best compaction of the soil

MEE pipe pipe that has been milled on each end and left rough in the center; MEE stands for "milled each end"

membrane roofing built-up roofing

mesh common term for welded wire fabric, plaster lath

mil 0.001"

military specifications these are specifications that the military writes for the products it buys from the manufacturers

minute 1/60th of a degree

MOA pipe pipe that has been milled end to end; MOA stands for "milled over all" and allows easier joining of the pipe if the length must be cut to fit

moisture barrier a material used for the purpose of resisting exterior moisture penetration

moisture density curve a graph plotted from tests to determine at what point of added moisture the maximum density will occur

moldings trim mounted around windows, floors, and doors as well as closets

monolithic concrete concrete placed as a single unit including turndown footings

mortar a concrete mix especially used for bonding masonry units

N

natural grade existing or original grade elevation of a property

natural ground the original ground elevation before any excavation has been done

nominal size original cut size of a piece of lumber prior to milling and drying; size of masonry unit, including mortar bed and head joint

non-bearing not supporting any structural load

nuclear test a test to determine soil compaction by sending nuclear impulses into the compacted soil and measuring the returned impulses reflected from the compacted particles

O

on center (O/C) the distance between the centers of two adjacent components

open web joist roof joist made of wood or steel construction with a top chord and bottom chord connected by diagonal braces bolted or welded together

P

package air conditioner or boiler an air conditioner or boiler in which all components are packaged into a single unit

pad in earthwork or concrete foundation work, the base materials used upon which to place the concrete footing and/or slab

parapet an extension of an exterior wall above the line of the roof

parging a thin moisture protection coating of plaster or mortar over a masonry wall

partition an interior wall separating two rooms or areas of building; usually non-bearing

penny (d) the unit of measure of the nails used by carpenters

perimeter the outside edges of a plot of land or building; it represents the sum of all the individual sides

perimeter insulation insulation placed around the outside edges of a slab

pile a steel or wooden pole driven into the ground sufficiently to support the weight of a wall and building

pillar a pole or reinforced wall section used to support the floor and consequently the building

pitch the slant or slope from the ridge to the plate

GLOSSARY (cont.)

plan view a bird's-eye view of a construction layout cut at 5'0" above finish floor level

plaster a mixture of cement, water and sand

plate a roof member which has the rafters fastened to it at their lower ends

platform framing also known as *western framing*; structural construction in which all studs are only one story high with joists over

point of beginning (POB) the point on a property from which all measurements and azimuths are established

polyvinyl chloride (PVC) a plastic material commonly used for pipe and plumbing fixtures

Portland cement one variety of cement produced from burning various materials such as clay, shale, and limestone, producing a fine gray powder; the basis of concrete and mortar

post-and-beam construction a type of wood frame construction using timber for the structural support

post-tensioning the application of stretching steel cables embedded in a concrete slab to aid in strengthening the concrete

prehung refers to doors or windows that are already mounted in a frame and are ready for installation as a complete unit

pressure treatment impregnating lumber with a preservative chemical under pressure in a tank

primer the first coat of paint or glue when more than one coat will be applied

purlin a horizontal framing member spanning between rafters

Q

quarry tile an unglazed clay or shale flooring material produced by the extrusion process

quick set a fast-curing cement plaster

R

R factor the numerical rating given any material that is able to resist heat transfer for a specific period of time

R values the unit that measures the effectiveness of insulation; the higher the number, the better the insulation qualities of the materials

rabbet a groove cut in or near the edge of a piece of lumber to fit the edge of another piece

raceway any partially or totally enclosed container for placing electrical wires (conduit, and so on)

rafter in sloped roof construction, the framing member extending from the ridge or hip to the top plate

rebar a reinforcement steel rod in a concrete footing

resilient flooring flooring made of plastics rather than wood products

ridge the highest point on a sloped roof

ridge board a horizontal member that connects the upper ends of the rafters on one side to the rafters on the opposite side

right-of-way line a line on the side of a road marking the limit of the construction area and usually, the beginning of private property

rise in roofing, rise is the vertical distance between the top of the double plate and the center of the ridge board; in stairs, it is the vertical distance from the top of a stair tread to the top of the next tread

riser the vertical part at the edge of a stair

roll roofing a type of built-up roofing material made of a mixture of rag, paper and asphalt

roof pitch the ratio of total span to total rise expressed as a fraction

rough opening a large opening made in a wall frame or roof frame to allow the insertion of a door or window

RS reference stake, from which measurements and grades are established

run the run of a roof is the shortest horizontal distance measured from a plumb line through the center of the ridge to the outer edge of the plate

S

sand cone test a test for determining the compaction level of soil, by removing an unknown quantity of soil and replacing it with a known quantity of sand

scabs boards used to join the ends of a girder

schematic a one-line drawing for electrical circuitry or isometric plumbing diagrams

scissors truss a truss constructed to the roof slope at the top chord with the bottom chord designed with a lower slope for interior vaulted or cathedral ceilings

scraper a digging, hauling, and grading machine having a cutting edge, a carrying bowl, a movable front wall, and a dumping mechanism

scratch coat first coat of plaster placed over lath in a three-coat plaster system

scupper an opening in a parapet wall attached to a downspout for water drainage from the roof

scuttle attic or roof access with cover or door

sealant a material used to seal off openings against moisture and air penetration

section a vertical drawing showing architectural or structural interior design developed at the point of a cutting-plane line on a plan view; the section may be transverse — the gable end — or longitudinal — parallel to the ridge

seismic design construction designed to withstand earthquakes

septic system a waste system used in lieu of a sewer system that includes a line from the structure to a tank and a leach field

shakes shingles made of handsplit wood; in most cases western cedar

shear wall a wall construction designed to withstand shear pressure caused by wind or earthquake

sheathing the outside layer of wood applied to studs to close up a house or wall; also used to cover the rafters and make a base for the roofing

sheepsfoot roller a compacting roller with feet expanded at their outer tips; used in compacting soil

shoring temporary support made of metal or wood used to support other components

sill a piece of wood that is anchored to the foundation

sinker nail a nail for laying subflooring; the head is sloped toward the shank but is flat on top

size size is a special coating used for walls before wallpaper is applied; it seals the walls and allows the wallpaper paste to attach itself to the wall and paper without adding undue moisture to the wall

slab-on-grade the foundation construction for a structure with no crawl space or basement

slump the consistency of concrete at the time of placement

soffit a covering for the underside of the overhang of a roof

soleplate a 2 x 4 or 2 x 6 used to support studs in a horizontal position; it is placed against the flooring and nailed into position onto the subflooring

span the horizontal distance between exterior bearing walls in a transverse section

specifications the written instructions detailing the requirements of construction for a project

spoil site area used to dispose of unsuitable or excess excavation material

spreader braces used across the top of concrete forms

square refers to a roof-covering area; a square consists of 100 square feet of area

stain a paint-like material that imparts a color to wood

stepped footing a footing that may be located on a number of levels

stool the flat shelf that rims the bottom of a window frame on the inside of a wall

stress skin panels large prebuilt panels used as walls, floors, and roof decks; built in a factory and hauled to the building site

string line a nylon line usually strung tightly between supports to indicate both direction and elevation; used in checking grades or deviations in slopes or rises

strip flooring wooden strips that are applied perpendicular to the joists

strongbacks braces used across ceiling joists that help align, space, and strengthen joists for drywall installation

structural steel heavy steel members larger than 12 gauge identified by their shapes

stucco a type of finish used on the outside of a building; a masonry finish that can be put on over any type of wall; applied over a wire mesh nailed to the wall

stud the vertical boards (usually 2 x 4 or 2 x 6) that make up the walls of a building

subfloor a platform that supports the rest of the structure; also referred to as underlayment

subgrade the uppermost level of material placed in embankment or left at cuts in the normal grading of a road bed

summit the highest point of any area or grade

super a continuous slope in one direction on a road

swale a shallow dip made to allow the passage of water

sway brace a piece of 2 x 4 or similar material used to temporarily brace a wall from wind until it is secured

swedes a method of setting grades at a center point by sighting across the tops of three lath; two lath are placed at a known correct elevation and the third is adjusted until it is at the correct elevation

symbol a pictorial representation of a material or component on a plan

T

tail/rafter tail that portion of a roof rafter extending beyond the plate line

tail joist a short beam or joist supported in a wall on one end and by a header on the other

tamp to pack tightly; usually refers to making sand tightly packed or making concrete mixed properly in a form to get rid of air pockets that may form with a quick pouring

tangent a straight line from one point to another, which passes over the edge of a curve

taping and bedding refers to drywall finishing; taping is the application of a strip of specially prepared tape to drywall joints; bedding means embedding the tape in the joint to increase its structural strength

tensile strength the maximum stretching of a piece of metal (rebar and so on) before breaking; calculated in kps

tensioning pulling or stretching of steel tendons to aid in reinforcement of concrete

terrazzo a mixture of concrete, crushed stone, calcium shells, and/or glass, polished to form a tile-like finish

texture paint a very thick paint that will leave a texture or pattern; can be shaped to cover cracked ceilings or walls or beautify an otherwise dull room

thermal ceilings ceilings that are insulated with batts of insulation to prevent loss of heat or cooling

tie a soft metal wire that is twisted around a rebar or reinforcement rod and chair to hold the rod in place till concrete is poured

tied out the process of determining the fixed location of existing objects (manholes, meter boxes, etc.) in a street so that they may be uncovered and raised after paving

toe of slope the bottom of an incline

top chord the topmost member of a truss

top plate the horizontal framing member fastened to the top of the wall studs; usually doubled

GLOSSARY (cont.)

track loader a loader on tracks used for filling and loading materials

tread the part of a stair on which people step

trimmer a piece of lumber, usually a 2 x 4, that is shorter than the stud or rafter but is used to fill in where the longer piece would have been normally spaced except for the window or door opening or some other opening in the roof or floor or wall

truss a prefabricated sloped roof system incorporating a top chord, bottom chord, and bracing

U

underlayment also referred to as the subfloor; used to support the rest of the building; may also refer to the sheathing used to cover rafters and serve as a base for roofing

unfaced insulation insulation which does not have a facing or plastic membrane over one side of it; it has to be placed on top of existing insulation; if used in a wall, it has to be covered by a plastic film to ensure a vapor barrier

V

valley the area of a roof where two sections come together and form a depression

valley rafters a rafter which extends diagonally from the plate to the ridge at the line of intersection of two roof surfaces

vapor barrier the same as a moisture barrier

veneer a thin layer or sheet of wood

vent usually a hole in the eaves or soffit to allow the circulation of air over an insulated ceiling; usually covered with a piece of metal or screen

vent stack a system of pipes used for air circulation and prevent water from being suctioned from the traps in the waste disposal system

ventilation the exchange of air, or the movement of air through a building; may be done naturally through doors and windows or mechanically by motor-driven fans

vibratory roller a self-powered or towed compacting device which mechanically vibrates while it rolls

GLOSSARY (cont.)

W

waler a 2x piece of lumber installed horizontally to formwork to give added stability and strength to the forms

water-cement ratio the ratio between the weight of water to cement

waterproofing preferably called moisture protection; materials used to protect below- and on-grade construction from moisture penetration

water table the amount of water that is present in any area; the moisture may be from rain or snow

welded wire fabric (WWF) a reinforcement used for horizontal concrete strengthening

wind lift (wind load) the force exerted by the wind against a structure caused by the movement of the air

winder fan-shaped steps that allow the stairway to change direction without a landing

window apron the flat part of the interior trim of a window; located next to the wall and directly beneath the window stool

window stool the flat narrow shelf which forms the top member of the interior trim at the bottom of a window

windrow the spill-off from the ends of a dozer or grader blade which forms a ridge of loose material; a windrow may be deliberately placed for spreading by another machine

wythe a continuous masonry wall width

XYZ

x brace cross brace for joist construction

zinc non-corrosive metal used for galvanizing other metals

ABBREVIATIONS

A	area	CC	center to center, cubic centimeter
AB	anchor bolt	CEM	cement
AC	alternate current	CER	ceramic
A/C	air conditioning	CFM	cubic feet per minute
ACT	acoustical ceiling tile	CIP	cast-in-place, concrete-in-place
A.F.F.	above finish floor		
AGGR	aggregate	CJ	ceiling joist, control joint
AIA	American Institute of Architects	CKT	circuit (electrical)
		CLG	ceiling
AL, ALUM	aluminum	CMU	concrete masonry unit
AMP	ampere		
APPROX	approximate	CO	cleanout (plumbing)
ASPH	asphalt	COL	column
ASTM	American Society for Testing Materials	CONC	concrete
		CONST	construction
		CONTR	contractor
AWG	American wire gauge	CU FT (ft³)	cubic foot (feet)
		CU IN (in³)	cubic inch(es)
BD	board	CU YD (yd³)	cubic yard(s)
BD FT (BF)	board foot (feet)		
BLDG	building	d	pennyweight (nail)
BLK	black, block		
BLKG	blocking	DC	direct current (elec.)
BM	board measure		
		DET	detail

ABBREVIATIONS (cont.)

DIA	diameter	FOS	face of stud, flush on slab
DIAG	diagonal	FT	foot, feet
DIM	dimension	FTG	footing
DN	down	FURN	furnishing, furnace
DO	ditto (same as)		
DS	downspout	FX GL (FX)	fixed glass
DWG	drawing		
		GA	gauge
E	East	GAL	gallon
EA	each	GALV	galvanize(d)
ELEC	electric, electrical	GD	ground (earth/electric)
ELEV	elevation, elevator		
		GI	galvanized iron
ENCL	enclosure	GL	glass
EXCAV	excavate, excavation	GL BLK	glass block
		GLB, GLU-LAM	glue-laminated beam
EXT	exterior		
		GRD	grade, ground
FDN	foundation	GWB	gypsum wall board
FIN	finish		
FIN FLR	finish floor	GYP	gypsum
FIN GRD	finish grade		
FL, FLR	floor	HB	hose bibb
FLG	flooring	HDR	header
FLUOR	fluorescent	HDW	hardware
FOB	free-on-board, factory-on-board	HGT/HT	height
		HM	hollow metal
FOM	face of masonry	HORIZ	horizontal

ABBREVIATIONS *(cont.)*

HP	horsepower	MAT'L	material
HWH	hot water heater	MAX	maximum
		MBF/MBM	thousand board feet, thousand board measure
ID	inside diameter		
IN	inch(es)	MECH	mechanical
INSUL	insulation	MISC	miscellaneous
INT	interior	MK	mark (identifier)
		MO	momentary (electrical contact), masonry opening
J, JST	joist		
JT	joint		
KG	kilogram		
KL	kiloliter	N	North
KM	kilometer	NEC	National Electric Code
KWH	kilowatt-hour		
		NIC	not in contract
L	left, line	NOM	nominal
LAU	laundry		
LAV	lavatory	O/A	overall (measure)
LBR	labor	O.C. (O/C)	on center
LDG	landing, leading	OD	outside diameter
LDR	leader	OH	overhead
LEV/LVL	level	O/H	overhang (eave line)
LIN FT (LF)	lineal foot (feet)		
LGTH	length	OPG	opening
LH	left hand	OPP	opposite
LITE/LT	light (window pane)		

ABBREVIATIONS *(cont.)*

PC	piece	S	South
PLAS	plastic	SCH/SCHED	schedule
PLAST	plaster	SECT	section
PLT	plate (framing)	SERV	service (utility)
PR	pair	SEW	sewer
PREFAB	prefabri-cate(d)(tion)	SHTHG	sheathing
		SIM	similar
PTN	partition	SP	soil pipe (plumbing)
PVC	polyvinylchloride pipe		
		SPEC	specification
		SQ FT (ft²)	square foot (feet)
QT	quart		
QTY	quantity	SQ IN (in²)	inch(es)
		SQ YD (yd²)	square yard(s)
R	right	STA	station
RD	road, round, roof drain	STD	standard
		STIR	stirrup (rebar)
REBAR	reinforced steel bar	STL	steel
		STR/ST	street
RECEPT	receptacle	STRUCT	structural
REINF	reinforce(ment)	SUSP CLG	suspended ceiling
REQ'D	required		
RET	retain(ing), return	SYM	symbol, symmetric
RF	roof	SYS	system
RFG	roofing (materials)		
RH	right hand		

ABBREVIATIONS (cont.)

T&G	tongue and groove	USE	underground service entrance cable (electrical)
THK	thick		
TOB	top of beam	W	West
TOC	top of curb	w/	with
TOF	top of footing	w/o	without
TOL	top of ledger	WC	water closet (toilet)
TOP	top of parapet		
TOS	top of steel	WDW	window
TR	tread, transition	WI	wrought iron
TRK	track, truck	WP	waterproof, weatherproof
TYP	typical		
		WT/WGT	weight
UF	underground feeder (electrical)	YD	yard
		Z	zinc

SYMBOLS

&	and	"	ditto, inch, -es
∠	angle	'	foot, feet
@	at	%	percent
#	number, pound	ø	diameter

CHAPTER 11
Conversion Factors

COMMONLY USED CONVERSION FACTORS		
Multiply	**By**	**To Obtain**
Acres...............	43,560	Square feet
Acres...............	1.562×10^{-3}	Square miles
Acre-Feet	43,560	Cubic feet
Amperes per sq. cm.	6.452	Amperes per sq. in.
Amperes per sq. in.	0.1550	Amperes per sq. cm.
Ampere-Turns..........	1.257	Gilberts
Ampere-Turns per cm. ...	2.540	Ampere-turns per in.
Ampere-Turns per in.	0.3937	Ampere-turns per cm.
Atmospheres	76.0	Cm. of mercury
Atmospheres	29.92	Inches of mercury
Atmospheres	33.90	Feet of water
Atmospheres	14.70	Pounds per sq. in.
British thermal units	252.0	Calories
British thermal units	778.2	Foot-pounds
British thermal units	3.960×10^{-4}	Horsepower-hours
British thermal units	0.2520	Kilogram-calories
British thermal units	107.6	Kilogram-meters
British thermal units	2.931×10^{-4}	Kilowatt-hours
British thermal units	1,055	Watt-seconds
B.t.u. per hour	2.931×10^{-4}	Kilowatts
B.t.u. per minute........	2.359×10^{-2}	Horsepower
B.t.u. per minute........	1.759×10^{-2}	Kilowatts
Bushels..............	1.244	Cubic feet
Centimeters	0.3937	Inches
Circular mils	5.067×10^{-6}	Square centimeters
Circular mils	0.7854×10^{-6}	Square inches
Circular mils	0.7854	Square mils
Cords	128	Cubic feet
Cubic centimeters	6.102×10^{-6}	Cubic inches

COMMONLY USED CONVERSION FACTORS *(cont.)*

Multiply	By	To Obtain
Cubic feet............	0.02832	Cubic meters
Cubic feet............	7.481	Gallons
Cubic feet............	28.32	Liters
Cubic inches..........	16.39	Cubic centimeters
Cubic meters	35.31	Cubic feet
Cubic meters	1.308	Cubic yards
Cubic yards	0.7646	Cubic meters
Degrees (angle)........	0.01745	Radians
Dynes...............	2.248×10^{-6}	Pounds
Ergs................	1	Dyne-centimeters
Ergs................	7.37×10^{-6}	Foot-pounds
Ergs................	10^{-7}	Joules
Farads.............	10^{6}	Microfarads
Fathoms	6	Feet
Feet................	30.48	Centimeters
Feet of water	.08826	Inches of mercury
Feet of water	304.8	Kg. per square meter
Feet of water	62.43	Pounds per square ft.
Feet of water	0.4335	Pounds per square in.
Foot-pounds..........	1.285×10^{-2}	British thermal units
Foot-pounds..........	5.050×10^{-7}	Horsepower-hours
Foot-pounds..........	1.356	Joules
Foot-pounds..........	0.1383	Kilogram-meters
Foot-pounds..........	3.766×10^{-7}	Kilowatt-hours
Gallons	0.1337	Cubic feet
Gallons	231	Cubic inches
Gallons	3.785×10^{-3}	Cubic meters
Gallons	3.785	Liters
Gallons per minute......	2.228×10^{-3}	Cubic feet per sec.
Gausses	6.452	Lines per square in.
Gilberts.............	0.7958	Ampere-turns
Henries	10^{3}	Millihenries
Horsepower	42.41	B.t.u. per min.
Horsepower	2,544	B.t.u. per hour

COMMONLY USED CONVERSION FACTORS *(cont.)*

Multiply	By	To Obtain
Horsepower	550	Foot-pounds per sec.
Horsepower	33,000	Foot-pounds per min.
Horsepower	1.014	Horsepower (metric)
Horsepower	10.70	Kg. calories per min.
Horsepower	0.7457	Kilowatts
Horsepower (boiler)	33,520	B.t.u. per hour
Horsepower-hours	2,544	British thermal units
Horsepower-hours	1.98×10^6	Foot-pounds
Horsepower-hours	2.737×10^5	Kilogram-meters
Horsepower-hours	0.7457	Kilowatt-hours
Inches.	2.540	Centimeters
Inches of mercury.	1.133	Feet of water
Inches of mercury.	70.73	Pounds per square ft.
Inches of mercury.	0.4912	Pounds per square in.
Inches of water.	25.40	Kg. per square meter
Inches of water.	0.5781	Ounces per square in.
Inches of water.	5.204	Pounds per square ft
Joules.	9.478×10^{-4}	British thermal units
Joules.	0.2388	Calories
Joules.	10^7	Ergs
Joules.	0.7376	Foot-pounds
Joules.	2.778×10^{-7}	Kilowatt-hours
Joules.	0.1020	Kilogram-meters
Joules.	1	Watt-seconds
Kilograms	2.205	Pounds
Kilogram-calories	3.968	British thermal units
Kilogram meters	7.233	Foot-pounds
Kg per square meter	3.281×10^{-3}	Feet of water
Kg per square meter	0.2048	Pounds per square ft.
Kg per square meter	1.422×10^{-3}	Pounds per square in.
Kilolines	10^3	Maxwells
Kilometers.	3.281	Feet
Kilometers.	0.6214	Miles
Kilowatts.	56.87	B.t.u. per min.

COMMONLY USED CONVERSION FACTORS (cont.)

Multiply	By	To Obtain
Kilowatts.	737.6	Foot-pounds per sec.
Kilowatts.	1.341	Horsepower
Kilowatts-hours	3409.5	British thermal units
Kilowatts-hours	2.655×10^6	Foot-pounds
Knots	1.152	Miles
Liters.	0.03531	Cubic feet
Liters.	61.02	Cubic inches
Liters.	0.2642	Gallons
Log N_e or in N	0.4343	Log_{10} N
Log N	2.303	Log_e N or in N
Lumens per square ft. . . .	1	Footcandles
Maxwells.	10^{-3}	Kilolines
Megalines	10^6	Maxwells
Megaohms	10^6	Ohms
Meters.	3.281	Feet
Meters.	39.37	Inches
Meter-kilograms	7.233	Pound-feet
Microfarads.	10^{-6}	Farads
Microhms	10^{-6}	Ohms
Microhms per cm. cube . .	0.3937	Microhms per in. cube
Microhms per cm. cube . .	6.015	Ohms per mil. foot
Miles.	5,280	Feet
Miles.	1.609	Kilometers
Miner's inches.	1.5	Cubic feet per min.
Ohms	10^{-6}	Megohms
Ohms	10^6	Microhms
Ohms per mil foot.	0.1662	Microhms per cm. cube
Ohms per mil foot.	0.06524	Microhms per in. cube
Poundals.	0.03108	Pounds
Pounds	32.17	Poundals
Pound-feet	0.1383	Meter-Kilograms
Pounds of water	0.01602	Cubic feet
Pounds of water	0.1198	Gallons
Pounds per cubic foot . . .	16.02	Kg. per cubic meter
Pounds per cubic foot . . .	5.787×10^{-4}	Pounds per cubic in.

COMMONLY USED CONVERSION FACTORS *(cont.)*

Multiply	By	To Obtain
Pounds per cubic inch . . .	27.68	Grams per cubic cm.
Pounds per cubic inch . . .	2.768×10^{-4}	Kg. per cubic meter
Pounds per cubic inch . . .	1.728	Pounds per cubic ft.
Pounds per square foot . .	0.01602	Feet of water
Pounds per square foot . .	4.882	Kg. per square meter
Pounds per square foot . .	6.944×10^{-3}	Pounds per sq. in.
Pounds per square inch . .	2.307	Feet of water
Pounds per square inch . .	2.036	Inches of mercury
Pounds per square inch . .	703.1	Kg. per square meter
Radians.	57.30	Degrees
Square centimeters	1.973×10^{5}	Circular mils
Square Feet	2.296×10^{-5}	Acres
Square Feet	0.09290	Square meters
Square inches.	1.273×10^{6}	Circular mils
Square inches.	6.452	Square centimeters
Square Kilometers	0.3861	Square miles
Square meters	10.76	Square feet
Square miles.	640	Acres
Square miles.	2.590	Square kilometers
Square Millimeters	1.973×10^{3}	Circular mils
Square mils.	1.273	Circular mils
Tons (long)	2,240	Pounds
Tons (metric).	2,205	Pounds
Tons (short).	2,000	Pounds
Watts	0.05686	B.t.u. per minute
Watts	10^{7}	Ergs per sec.
Watts	44.26	Foot-pounds per min.
Watts	1.341×10^{-3}	Horsepower
Watts	14.34	Calories per min.
Watts-hours	3.412	British thermal units
Watts-hours	2,655	Footpounds
Watts-hours	1.341×10^{-3}	Horsepower-hours
Watts-hours	0.8605	Kilogram-calories
Watts-hours	376.1	Kilogram-meters
Webers	10^{8}	Maxwells

CONVERSION TABLE FOR TEMPERATURE – °F / °C

°F	°C	°F	°C	°F	°C	°F	°C	°F	°C
-459.4	-273	-22.0	-30	35.6	.2	93.2	.34	150.8	.66
-418.0	-250	-18.4	-28	39.2	.4	96	.36	154.4	.68
-328.0	-200	-14.8	-26	42.8	.6	100.4	.38	158.0	.70
-238.0	-150	-11.2	-24	46.4	.8	104.0	.40	161.6	.72
-193.0	-125	-7.6	-22	50.0	.10	107.6	.42	165.2	.74
-148.0	-100	-4.0	-20	53.6	.12	111.2	.44	168.8	.76
-130.0	-90	-0.4	-18	57.2	.14	114.8	.46	172.4	.78
-112.0	-80	3.2	-16	60.8	.16	118.4	.48	176.0	.80
-94.0	-70	6.8	-14	64.4	.18	122.0	.50	179.6	.82
-76.0	-60	10.4	-12	68.0	.20	125.6	.52	183.2	.84
-58.0	-50	14.0	-10	71.6	.22	129.2	.54	186.8	.86
-40.0	-40	17.6	-8	75.2	.24	132.8	.56	190.4	.88
-36.4	-38	21.2	-6	78.8	.26	136.4	.58	194.0	.90
-32.8	-36	24.8	-4	82.4	.28	140.0	.60	197.6	.92
-29.2	-34	28.4	-2	86.0	.30	143.6	.62	201.2	.94
-25.6	-32	32.0	.0	89.6	.32	147.2	.64	204.8	.96

°F	°C	°F	°C	°F	°C	°F	°C	°F	°C
208.4	.98	347.0	175	590	310	1004	540	6332	3500
212.0	100	356.0	180	608	320	1040	560	7232	4000
221.0	105	365.0	185	626	330	1076	580	4500	8132
230.0	110	374.0	190	644	340	1112	600	9032	5000
239.0	115	383.0	195	662	350	1202	650	9932	5500
248.0	120	392.0	200	680	360	1292	700	10832	6000
257.0	125	410	210	698	370	1382	750	11732	6500
266.0	130	428	220	716	380	1472	800	12632	7000
275.0	135	446	230	734	390	1562	850	13532	7500
284.0	140	464	240	752	400	1652	900	14432	8000
293.0	145	482	250	788	420	1742	950	15332	8500
302.0	150	500	260	824	440	1832	1000	16232	9000
311.0	155	518	270	860	460	2732	1500	17132	9500
320.0	160	536	280	896	480	3632	2000	18032	10000
329.0	165	554	290	932	500	4532	2500		
338.0	170	572	300	968	520	5432	3000		

1 degree F is 1/180 of the difference between the temperature of melting ice and boiling water.
1 degree C is 1/100 of the difference between the temperature of melting ice and boiling water.
Absolute Zero = 273.16°C = -459.69° F

DECIBEL LEVELS OF SOUNDS

The definition of sound intensity is energy (erg) transmitted per 1 second over a square centimeter surface. Sounds are measured in decibels. A decibel (db) change of 1 is the smallest change detected by humans.

Hearing Intensity	Decibel Level	Examples of Sounds
Barely	0	Dead silence
Audible		Audible hearing threshold
	10	Room (sound proof)
(Very	20	Empty auditorium
Light)		Ticking of a stopwatch
		Soft whispering
Audible	30	People talking quietly
Light	40	Quiet street noise without autos
Medium	45	Telephone operator
Loud	50	Fax machine in office
	60	Close conversation
Loud	70	Stereo system
		Computer printer
	80	Fire truck/ambulance siren
		Cat/dog fight
Extremely	90	Industrial machinery
Loud		High school marching band
Damage	100	Heavy duty grinder in a
Possible		machine/welding shop
Damaging	100+	Begins ear damage
	110	Diesel engine of a train
	120	Lightning strike (thunderstorm)
		60 ton metal forming factory press
	130	60" fan in a bus vacuum system
	140	Commercial/military jet engine
Ear Drum	194	Space shuttle engines
Shattering	225	16" guns on a battleship

SOUND AWARENESS AND SAFETY

Sound Awareness Changes

The typical range of human hearing is 30 hertz - 15,000 hertz. Human hearing recognizes an increase of 20 decibels, such as a stereo sound level increase, as being four times as loud at the higher level than it was at the lower level.

Awareness in Human Hearing	Decibel Change
Noticeably Louder	10
Easily Audible	5
Faintly Audible	3

HEARING PROTECTION LEVELS

Because of the Occupational Safety and Health Act of 1970, hearing protection is mandatory if the following time exposures to decibel levels are exceeded because of possible damage to human hearing.

Decibel Level	Time Exposure Per Day
115	15 minutes
110	30 minutes
105	1 hour
102	1½ hours
100	2 hours
97	3 hours
95	4 hours
92	6 hours
90	8 hours

TRIGONOMETRIC FORMULAS – RIGHT TRIANGLE

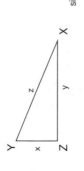

Angles = X, Y, Z
Distances = x, y, z
Area = $\dfrac{x\,y}{2}$

Pythagorean Theorem states
That $x^2 + y^2 = z^2$

Thus $x = \sqrt{z^2 - y^2}$

Thus $y = \sqrt{z^2 - x^2}$

Thus $z = \sqrt{x^2 + y^2}$

$\sin X = \dfrac{x}{z}$ $\cos X = \dfrac{y}{z}$

$\tan X = \dfrac{x}{y}$ $\cot X = \dfrac{y}{x}$

Given x and z, find X, Y and y

$\sin X = \dfrac{x}{z} = \cos Y, \; y = \sqrt{(z^2 - x^2)} = z\sqrt{1 - \dfrac{x^2}{z^2}}$

Given x and y, find X, Y and z

$\tan X = \dfrac{x}{y} = \cot Y, \; z = \sqrt{x^2 + y^2} = x\sqrt{1 + \dfrac{y^2}{x^2}}$

Given X and z, find Y, x and y

$Y = 90^\circ - X, \; x = z \sin X, \; y = z \cos X$

Given X and z, find Y, x and z

$Y = 90^\circ - X, \; x = y \tan X, \; z = \dfrac{y}{\cos X}$

Given X and x, find Y, y and z

$Y = 90^\circ - X, \; y = x \cot X, \; z = \dfrac{x}{\sin X}$

TRIGONOMETRIC FORMULAS – OBLIQUE TRIANGLES

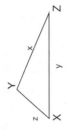

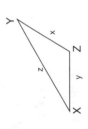

Given x, y and z, Find X, Y and Z

$$s = \frac{x+y+z}{2}, \quad \sin\tfrac{1}{2}X = \sqrt{\frac{(s-y)(s-z)}{yz}}$$

$$\sin\tfrac{1}{2}Y = \sqrt{\frac{(s-x)(s-z)}{xz}}, \quad C = 180^\circ - (X+Y)$$

Given x, y and z, find the Area

$$s = \frac{x+y+z}{2}, \quad \text{Area} = \sqrt{S(s-x)(s-y)(s-z)}$$

$$\text{Area} = \frac{yz \sin X}{2}, \quad \text{Area} = \frac{x^2 \sin Y \sin Z}{2 \sin X}$$

Given x, y, and Z, find X, Y and z

$$X + Y = 180^\circ - Z, \quad z = \frac{x \sin Z}{\sin X}, \quad \tan X = \frac{x \sin Z}{y - (x \cos Z)}$$

Given X, x and y, Find Y, Z and z

$$\sin Y = \frac{y \sin X}{x}, \quad Z = 180^\circ - (X+Y), \quad z = \frac{x \sin Z}{\sin X}$$

Given X, Y and x, Find y, Z and z

$$y = \frac{x \sin Y}{\sin X}, \quad Z = 180^\circ - (X+Y), \quad z = \frac{x \sin Z}{\sin X}$$

TRIGONOMETRIC FORMULAS – SHAPES

Equilateral Triangle	Annulus	Trapezium

Equilateral Triangle

X = Sides (Equal Lengths)

$Area = X^2 \sqrt{\dfrac{3}{4}} = .433 X^2$

$Perimeter = 3 X$

$H = \dfrac{X}{2} \sqrt{3} = .866 X$

Annulus

C_1 and R_1 = Inside Circle

C_2 and R_2 = Outside Circle

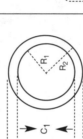

C = Circumference

R = Radius

$Area = \pi (R_1 + R_2)(R_2 - R_1)$

$Area = \left((C_2)^2 - (C_1)^2 \right) .7854$

Trapezium

Perimeter is the

Sum of L, M, N and O

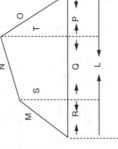

$Area = \dfrac{(S + T) Q + RS + PT}{2}$

11-12

Trapezoid

Perimeter =
The Sum of the
lengths of all
four sides

$$Area = \frac{(X + Y)}{2}$$

Quadrilateral

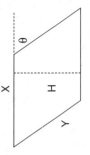

$$Area = \frac{L_1 \bullet L_2 \bullet Sin\,\theta}{2}$$

Where θ =
Degrees
of Angle

Rectangle

$Area = XY$
Diagonal Line (D)

$$= \sqrt{X^2 + Y^2}$$

Perimeter =
$2(X + Y)$
If a square
then X = Y

Parallelogram

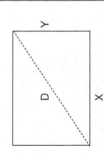

Where θ =
Degrees
of Angle

$Area =$
$XH = XY\,sin\,\theta$
Perimeter =
$2(X + Y)$

COMMON ENGINEERING UNITS AND THEIR RELATIONSHIP

Quantity	SI Metric Units/Symbols	Customary Units	Relationship of Units
Acceleration	meters per second squared (m/s^2)	feet per second squared (ft/s^2)	$m/s^2 = ft/s^2 \times 3.281$
Area	square meter (m^2) square millimeter (mm^2)	square foot (ft^2) square inch (in^2)	$m^2 = ft^2 \times 10.764$ $mm^2 = in^2 \times 0.00155$
Density	kilograms per cubic meter (kg/m^3) grams per cubic centimeter (g/cm^3)	pounds per cubic foot (lb/ft^3) pounds per cubic inch (lb/in^3)	$kg/m^3 = lb/ft^3 \times 16.02$ $g/cm^3 = lb/in^3 \times 0.036$
Work	Joule (J)	foot pound force (ft lbf or ft lb)	$J = ft\ lbf \times 1.356$
Heat	Joule (J)	British thermal unit (Btu) Calorie (Cal)	$J = Btu \times 1.055$ $J = cal \times 4.187$
Energy	kilowatt (kW)	Horsepower (HP)	$kW = HP \times 0.7457$

11-14

		Pound-force (lbf, lb · f, or lb) kilogram-force (kgf, kg · f, or kp)	
Force	Newton (N) Newton (N)		$N = lbf \times 4.448$ $N = \dfrac{kgf}{9.807}$
Length	meter (m) millimeter (mm)	foot (ft) inch (in)	$m = ft \times 3.281$ $mm = \dfrac{in}{25.4}$
Mass	kilogram (kg) gram (g)	pound (lb) ounce (oz)	$kg = lb \times 2.2$ $g = \dfrac{oz}{28.35}$
Stress	Pascal = Newton per second (Pa = N/s)	pounds per square inch (lb/in^2 or psi)	$Pa = lb/in^2 \times 6{,}895$
Temperature	degree Celsius (°C)	degree Fahrenheit (F)	$°C = \dfrac{°F - 32}{1.8}$
Torque	Newton meter (N · m)	foot-pound (ft lb) inch-pound (in lb)	$N \cdot m = ft\ lbf \times 1.356$ $N \cdot m = in\ lbf \times 0.113$
Volume	cubic meter (m^3) cubic centimeter (cm^3)	cubic foot (ft^3) cubic inch (in^3)	$m^3 = ft^3 \times 35.314$ $cm^3 = \dfrac{in^3}{16.387}$

DECIMAL EQUIVALENTS OF FRACTIONS

8ths	32nds	64ths	64ths
1/8 = .125	1/32 = .03125	1/64 = 0.15625	33/64 = .515625
1/4 = .250	3/32 = .09375	3/64 = .046875	35/64 = .546875
3/8 = .375	5/32 = .15625	5/64 = .078125	37/64 = .57812
1/2 = .500	7/32 = .21875	7/64 = .109375	39/64 = .609375
5/8 = .625	9/32 = .28125	9/64 = .140625	41/64 = .640625
3/4 = .750	11/32 = .34375	11/64 = .171875	43/64 = .671875
7/8 = .875	13/32 = .40625	13/64 = .203128	45/64 = .703125
16ths	15/32 = .46875	15/64 = .234375	47/64 = .734375
1/16 = .0625	17/32 = .53125	17/64 = .265625	49/64 = .765625
3/16 = .1875	19/32 = .59375	19/64 = .296875	51/64 = .796875
5/16 = .3125	21/32 = .65625	21/64 = .328125	53/64 = .828125
7/16 = .4375	23/32 = .71875	23/64 = .359375	55/64 = .859375
9/16 = .5625	25/32 = .78125	25/64 = .390625	57/64 = .890625
11/16 = .6875	27/32 = .84375	27/64 = .421875	59/64 = .921875
13/16 = .8125	29/32 = .90625	29/64 = .453125	61/64 = .953125
15/16 = .9375	31/32 = .96875	31/64 = .484375	63/64 = .984375

COMMONLY USED GEOMETRICAL RELATIONSHIPS

Diameter of a circle × 3.1416 = Circumference.

Radius of a circle × 6.283185 = Circumference.

Square of the radius of a circle × 3.1416 = Area.

Square of the diameter of a circle × 0.7854 = Area.

Square of the circumference of a circle × 0.07958 = Area.

Half the circumference of a circle × half its diameter = Area.

Circumference of a circle × 0.159155 = Radius.

Square root of the area of a circle × 0.56419 = Radius.

Circumference of a circle × 0.31831 = Diameter.

Square root of the area of a circle × 1.12838 = Diameter.

Diameter of a circle × 0.866 = Side of an inscribed equilateral triangle.

Diameter of a circle × 0.7071 = Side of an inscribed square.

Circumference of a circle × 0.225 = Side of an inscribed square.

Circumference of a circle × 0.282 = Side of an equal square.

Diameter of a circle × 0.8862 = Side of an equal square.

Base of a triangle × one-half the altitude = Area.

Multiplying both diameters and .7854 together = Area of an ellipse.

Surface of a sphere × one-sixth of its diameter = Volume.

Circumference of a sphere × its diameter = Surface.

Square of the diameter of a sphere × 3.1416 = Surface.

Square of the circumference of a sphere × 0.3183 = Surface.

Cube of the diameter of a sphere × 0.5236 = Volume.

Cube of the circumference of a sphere × 0.016887 = Volume.

Radius of a sphere × 1.1547 = Side of an inscribed cube.

Diameter of a sphere divided by $\sqrt{3}$ = Side of an inscribed cube.

Area of its base × one-third of its altitude = Volume of a cone or
  pyramid whether round, square or triangular.

Area of one of its sides × 6 = Surface of the cube.

Altitude of trapezoid × one-half the sum of its parallel sides = Area.

About The Author

Paul Rosenberg has an extensive background in the construction, data, electrical, HVAC and plumbing trades. He is a leading voice in the electrical industry with years of experience from an apprentice to a project manager. Paul has written for all of the leading electrical and low voltage industry magazines and has authored more than 30 books.

In addition, he wrote the first standard for the installation of optical cables (ANSI-NEIS-301) and was awarded a patent for a power transmission module. Paul currently serves as contributing editor for *Power Outlet Magazine*, teaches for Iowa State University and works as a consultant and expert witness in legal cases. He speaks occasionally at industry events.